BIOGEOGRAPHY an ecological and evolutionary approach

C. BARRY COX PhD

IAN N. HEALEY PhD

PETER D. MOORE PhD

School of Biological Sciences
King's College, London

MANSFIELD COLLEGE
LIBRARY
WITHDRAWN
OXFORD

KU-785-998

BLACKWELL SCIENTIFIC PUBLICATIONS
OXFORD LONDON EDINBURGH MELBOURNE

© 1973 Blackwell Scientific Publications
Osney Mead, Oxford
3 Nottingham Street, London W1
9 Forrest Road, Edinburgh
P.O. Box 9, North Balwyn, Victoria, Australia

ISBN 0 632 09140 1 paper

First published 1973

Printed in Great Britain by
Burgess & Son Ltd.
Abingdon, Berkshire
and bound by
Kemp Hall Bindery
Osney Mead, Oxford

BIOGEOGRAPHY an ecological
and evolutionary approach

CONTENTS

PREFACE

Biogeography lies at the meeting-point of several fields of study. To understand why a particular group of organisms is found in a specific area requires knowledge of the organism's ecological relationships— why it is associated with this soil, this temperature range, or this type of woodland. To understand why it is found in particular areas of a continent may require knowledge of the climatic history, which may have led to the isolation of scattered relict communities. Finally, to understand why it is found in some continents and not in others involves knowledge of the evolutionary history of the group itself and of the geological histories of the land-masses, as the processes of continental drift transported them across the globe, splitting them asunder or welding them in new patterns.

As the impact of man upon the Earth's animals and plants increases dramatically, we have become more aware of the need to control the effects of our social and industrial habits. Biogeographers and conservationists therefore now also need to understand how man's present relationship to his environment has come about, and the ways in which he has altered, often radically, the structure of the ecosystems he inhabits or uses.

Despite biogeography's connections with ecology, geography, geology, evolutionary history and economic anthropology, it is usually studied in conjunction with only one or two of these disciplines. Our intention here has been to provide enough of the basic elements of all these approaches, so that a student who specializes in only one of them may nevertheless be able to understand the way in which all of them interact to produce the apparently bewildering array of patterns of distribution of life.

ACKNOWLEDGMENTS

We should like to acknowledge the assistance given by Aldus Books, and especially by Jonathan Elphick, in the preparation of the early stages of both the text and the illustrations of this book.

CHAPTER 1

PATTERNS OF LIFE

Life in all its physical and chemical complexity exists in a multitude of forms, or species. Current estimates show that about 300,000 species of green plants and fungi and about 1,300,000 species of animals have been recognized by biologists. These figures do not include the bacteria and yeasts, of which there are thousands of types, and undoubtedly there are many thousands of species of other groups of organisms remaining to be discovered. None of these forms of life is distributed haphazardly over the surface of the world. Each species occupies only a limited area of part of the world, although the size of the area occupied varies greatly from species to species. There are very rare species that are found in only one or two places, and others that are very common and found almost everywhere. But even the most common species—such as our own *Homo sapiens*—do not live everywhere; very few people live in the polar regions or in desert areas. In fact, unevenness of spatial distribution is as basic a characteristic of living organisms as locomotion or respiration.

Why should this be so? One reason is that during its history each species has evolved so that its life processes of physiology, growth, and behaviour function efficiently only within a limited range of environmental conditions, and with only certain types of food resources. This is probably a result of the pressures of competition between species for the limited space and food resources available in their environments. Only by ever-increasing specialization in the space it occupies and the food it uses can a species gain some competitive advantage over others. This process of specialization—adaptation to particular factors or combinations of factors in the environment—is a continuous one, and has occurred by the evolutionary process of natural selection, which will be discussed in a later chapter. For the present it can be assumed that because the physical conditions of the organism's environment—temperature, light, wetness or dryness, and so on—and the food resources it contains are far from evenly distributed, the distribution of organisms must also be uneven. Each species therefore has a pattern of distribution related to that of the physical conditions and food resources to which it is adapted.

What is biogeography? The study of the patterns of distribution of organisms in space and time is called *biogeography*. Biologists are nowadays rarely satisfied with the mere description of these patterns, and the biogeographer usually wants to discover which environmental factors are the ones that determine or limit the distribution of the species he studies. To do this he must draw on knowledge from the whole spectrum of the sciences of life and the environment, including geology, geophysics, climatology, palaeontology, plant and animal systematics and taxonomy, evolution, physiology, and ecology. This book is concerned with explaining the distribution of plants and animals rather than with merely cataloguing them. Its chief aim is to show how the physical environment of a species, its biology, and its evolutionary history interact to bring about its pattern of distribution. But we wish also to show that biogeography is not just an academic science, without relevance to human problems. In the future we shall have to find many new sources of food for our expanding population. Very probably we shall begin to use for food a number of species of animals and plants that we have not previously exploited, or at any rate to change drastically, by selective breeding programmes, the characters and productive capacity of species we already use. (This is already happening extensively with the development of new high-yielding varieties of rice for tropical countries and of varieties of maize or sweet corn that can grow in cold, wet countries like Britain.) The science of biogeography will often be able to predict whether the species we wish to grow or breed can survive and be productive in environments where they are not normally found, or whether we must develop varieties with new characters such as resistance to drought or cold.

Biogeography is important, too, in fields other than food production. Recently, people have at last begun to realize that the plants and animals that share our environment are a valuable resource, not only in the economic sense, but a resource also of interest and beauty which we have a duty to manage and conserve for future generations just like any other resource, such as food or energy. But the conservation and management of species of animals and plants in natural environments is a complex activity, much more difficult than the management of animals and plants in agriculture. It requires a detailed knowledge of their biology, especially of their geographical distribution, so that we may know in which environments they occur and what conditions they can tolerate. So biogeography is basic to programmes of conservation and management of environments.

As with other aspects of biology, there is no clear point at which biogeography ends and other related sciences, especially ecology, begin. This is because the distribution of a species can be studied on an infinite range of scales from the global to the local. This can be illustrated by reference to a mammal—the European badger, *Meles meles*[1]. On the global scale the distribution of the badger covers western and northern Europe as far as the Arctic Circle, Asia Minor, and a zone of central Asia from the Arctic Circle southward to the Himalayas, China, and Japan. Within this huge area of the earth's surface the badger is, of course, very far from evenly distributed. Its patchy distribution is related to a great many factors, but in general the badger is rarest in flat and marshy country and commonest in hilly, wooded areas. In the

British Isles, for instance, the badger is most common in southern and western England and rarest in East Anglia and parts of Ireland and Scotland. Where they occur, badgers most frequently make their burrows or *setts* in woods and copses, especially where these give easy access to the pastureland where badgers often feed, and where the soil is well-drained and suitable for digging. The biologist calls these places where the badger lives its *habitat*. Nearly all animal and plant species seem to have such recognizable habitats where they are found more often than in other places.

Habitats and microhabitats   For many organisms, especially larger ones, distribution can be conveniently considered in terms of such units of habitat as "woodland," "grassland," or "seashore." But most species have specific distributions even within such units of the environment as the habitat. The woodland habitat, for instance, consists of a host of smaller *microhabitats*—the humus and leaf-litter layer of the soil, rotting logs, the ground flora zone, the various levels of the tree canopy, tree trunks, and beneath the bark of living trees. Certain characteristic species of animals and plants are found in each of these microhabitats, and so the distribution of these species in woodland coincides more or less closely with that of their microhabitats. Some species are found in more than one of these areas, but generally each species has a particular microhabitat that may be termed its "headquarters," in which it occurs most frequently and in the highest numbers. Even within habitats that are simpler in structure than woodland—such as grassland—many different microhabitats occur and, as will be explained later, the number of microhabitats present is an important factor determining the number of species that may live in a habitat.

Many quite large and active animals show a tendency to confine themselves to certain parts of a large habitat. The spider monkeys (*Ateles*) of the lowland forests of central and northern South America are active climbers, able to jump long distances, and family groups move about the forest a great deal. If the monkeys are observed over a period of time, however, they can be seen to spend most of their time in the lower parts of the high canopy of the forest, and especially on the smaller peripheral branches of the trees. The reason for this is clear; over 90 per cent of the diet of spider monkeys consists of fruit and nuts, and these food resources will naturally be found most abundantly on the smaller, fast-growing parts of the trees. It is probably best to call areas like these, preferred by a largish, active animal, *minor habitats*, rather than microhabitats, and to keep the latter term for subdivisions with more clearly defined boundaries.

Limits of distribution   Surrounding the areas of a species' distribution, whether this is considered on a geographical, habitat, or microhabitat scale, are areas where the species cannot maintain a population because physical conditions or lack of food resources are too extreme to permit survival. These areas can be viewed as *barriers* that must be crossed by the species if it is to disperse to other favourable, but as yet uncolonized, places—much as the European settlers had to cross ocean barriers to colonize North America or Australia. Any climatic or topographic factor, or combination of factors, may provide a barrier to the distribution of an organism. For example, the problems of locomotion or of obtaining oxygen and food are quite different in water and air. As a result, organisms which are adapted for life on land are unable to cross oceans: their eventual death will be due, in varying proportions, to drowning, to starvation, to exhaustion and to lack of fresh water to drink. Similarly, land is a barrier to organisms which are adapted to life in sea or fresh water, because they require supplies of oxygen dissolved in water rather than as an atmospheric gas, and because they desiccate rapidly in air. Mountain ranges, too, form effective barriers to dispersal because they present extremes of cold too great for many organisms. The amount of rainfall, the rate of evaporation of water from the soil surface, and light intensity are all critical factors limiting the distribution of most green rooted plants. But in all these cases, and in most others, the ultimate barriers are not the hostile factors of the environment but the species' own physiology, which has become adapted to a limited range of environmental conditions. In its distribution a species is therefore the prisoner of its own evolutionary history.

At the habitat level, the microhabitats of organisms are surrounded by areas of small-scale variation of physical conditions, or *microclimates* —similar, but on a much smaller scale, to geographical variations in climate—and of food distribution. These form barriers restricting species to their microhabitats. The insects that live in rotting logs, for instance, are adapted by their evolution to a microhabitat with a high water content, and relatively constant temperatures. The logs provide the soft woody materials and micro-organisms they need for food, and also give good protection from predators. Around the logs are areas with fewer, or none, of these desirable qualities, and attempts by the animals to leave their microhabitat would result for many of them in death by desiccation, starvation, or predation.

Overcoming the barriers Nevertheless, a few inhabitants of rotting logs do occasionally make the dangerous journey from one log to another, and this shows that very few environmental factors are absolute barriers to the dispersal of organisms and that they vary greatly in their effectiveness. Most habitats and microhabitats have only limited resources, and the organisms living in them must have mechanisms enabling them to find new habitats and resources when the old ones become exhausted. These mechanisms often take the form of seeds, resistant stages, or—as in the case of the insects of the rotting-log microhabitat—flying adults with a fairly high resistance to desiccation.

There is good evidence that geographical barriers are not completely effective either. When organisms extend their distribution on a geographical scale it is likely that they are taking advantage of temporary, seasonal, or permanent changes of climate or distribution of habitats that allow them to cross barriers normally closed to them. The British Isles, for instance, lie within the geographical range of about 220 species of birds, but a further 50 or 60 species visit the region as so-called "accidentals"—these birds do not breed in Britain, but one or two individuals are seen by ornithologists every few years. They come for a variety of reasons: some are blown off-course by winds during migration, others are forced in certain years to leave their normal ranges when numbers are especially high and food is scarce. Many of these accidentals have their true home in North America—such as the pectoral sandpiper (*Calidris melanotos*), a few of which are seen every year—but some come from eastern Asia or even from the Pacific Islands. It is possible, though not very likely, that a few of these chance travellers may in time establish themselves permanently in Europe, as did the collared dove (*Streptopelia decaocto*) which in a few decades has spread from Asia Minor and southern Asia across central Europe and into the British Isles and Scandinavia—perhaps the most dramatic change in distribution known in any vertebrate.[2,3] This species is often common around the edges of towns and settlements and seems to depend for food largely on the seeds of weed species common in farms and gardens. Several factors may have interacted to permit this extension of range of the collared dove. Increased human activity during the last century, involving extensive changes in the environment, has produced new habitats and food resources, and it is possible, too, that small changes in climate may have significantly favoured this species. It is, however, considered unlikely that the collared dove would have been able to take advantage of these changes without a change in its own genetic make-up, perhaps a physiological one permitting the species to tolerate a wider range of climatic conditions or to utilize a wider range of food substances.[4]

The case of the collared dove shows how organisms can extend their distribution when new habitats become available. At the present time the human species has such an influence on the ecology of the whole world that new habitats are constantly being created and old ones destroyed. This causes rapid changes in the geographical distribution of many other species. In the past, new habitats became available at a slower rate, through climatic changes, fluctuations in sea level, glaciation, mountain formation, and erosion, but over long periods of time these slow changes have had a great effect on the distribution of plants and animals.

A discussion of some patterns of distribution shown by particular species of animals and plants will show how varied and complex these may be and will help to emphasize the various scales or levels on which such patterns may be considered. In fact the number of examples that we can choose is quite limited because the distribution of only a very small number of species has been investigated in detail. Even amongst well-known species, chance finds in unusual places are constantly modifying known distribution patterns and requiring changes in the explanations that biologists give of these patterns.

Some existing patterns are continuous, the area occupied by the group consisting of a single region or of a number of regions which are closely adjacent to one another. These patterns can usually be explained by the distribution of present-day climatic and biological factors; the detailed distributions of several species of dragon-fly, and of the plantains, provide good examples of these. Other existing patterns are discontinuous or "disjunct," the areas occupied being widely separated and scattered over a particular continent, or over the whole world. The organisms which show such a pattern may, like the cycads, be merely the scattered survivors of a once-dominant and widespread group, now unable to compete with newer forms. Others, the "climatic relicts" or "habitat relicts" appear to have been greatly affected by past changes in climate or sea level. Finally, as will be shown in Chapters 6 and 7, the disjunct patterns of some living groups, and of many extinct groups, have resulted from the physical splitting of a once-continuous area of distribution by the process of continental drift.

Patterns of dragonflies One group of species whose distributions are quite well known, at least in western Europe, are the Odonata, dragonflies and damselflies.[5] The common blue damselfly, *Enallagma cyathigerum*, is possibly one of the most abundant and widely distributed dragonfly species (Fig. 1a). In Britain the adults are on the wing in mid-summer, around bodies of fresh water. The female lays eggs in vegetation below the surface of the water and the larvae hatch in a week or two. These live on the bottom of the pond, stream or lake and feed on small crustaceans and insect larvae until they reach a size of 17–18 mm, which may take from two to four years, depending on the quality and quantity of food available. In the May or June after reaching full size, the larvae climb up the stems of *emergent vegetation* (plants rooted in the mud with stem and leaves emerging from the water surface), cast their larval skins, and emerge as winged adults. *E. cyathigerum* is found in a wide range of freshwater habitats, including both still and moving water, though it is perhaps least common in fast-moving water, or in places where silt is being deposited. Probably the ideal habitat for this species is a fairly large body of still water with plenty of floating vegetation, such as pondweed (*Potamogeton*), and with a good growth of shore-weed (*Littorella uniflora*) or water milfoil (*Myriophyllum alterniflorum*) on the bottom. None of these requirements is at all rigid, however, and big populations often occur elsewhere. *E. cyathigerum* is found in both acid and alkaline waters, and often occurs in brackish pools on salt marshes. An ecologist would describe this species as an *ecologically tolerant* or *eurytopic* one, because its habitat preferences are so broad. As a result of this tolerance it is found almost everywhere in the British

Figure 1. The distribution on a world scale and within the British Isles of three species of dragonfly. (a) *Enallagma cyathigerum*, a circumboreal distribution, (b) *Sympetrum sanguineum*, a limited distribution, and (c) *Anax imperator*, a disjunct distribution.

Isles, being absent only from Snowdonia, parts of the Pennines, and lowland and north-east Scotland.

The geographical distribution of this damselfly is also very wide. In Europe the distribution lies mostly between 45°N and the Arctic Circle, although it includes some of the wetter parts of Spain, and is rather scattered in northern Scandinavia; it is not found in Greenland or Iceland. The species is found in a few places in North Africa, in Asia Minor, and around the Caspian Sea. Large areas of Asia south of the Arctic Circle also fall within its range. In North America it is found everywhere north of about 35–40°N to the Arctic Circle, except Labrador and Baffin Island. Populations also occur in the ideal habitats provided by the Everglades of Florida, which mark the species' furthest southward expansion. The broad geographical distribution of this

7

species is almost certainly due to its ecological tolerance and ability to make use of a wide range of habitats in very different climates. This type of distribution pattern, a belt around the Northern Hemisphere, is shown by many species of animals and plants, and is termed *circumboreal*—"around the northern regions". The frequency of this pattern in very different organisms suggests that the two northern land-masses may once have been joined, enabling certain species to spread right around the hemisphere.

As *E. cyathigerum* is so successful one might ask why it has not spread further southward. One reason may be the relative scarcity of watery habitats in the subtropical regions immediately to the south of its present range—the arid areas of Central America, North Africa, and Central Asia. The species is perhaps not robust enough for the long migrations that would be needed to reach suitable habitats in the Southern Hemisphere (there are very few wind belts that might assist such a migration). Another possibility is that further south there are other species already occupying all the habitats that *E. cyathigerum* could colonize. These species may be better adapted to the physical conditions of their habitats than *E. cyathigerum* and could therefore compete successfully with it for the available food resources; this might exclude the species from these areas. In fact there are many species of the genus *Enallagma* and the closely related genus *Ischnura* in the Southern Hemisphere that might be expected to have similar habitat and food requirements to *E. cyathigerum*; there are at least eight species of *Enallagma* in South Africa alone. Any of these explanations, or a combination of them, would explain why the common blue damselfly is confined to the high latitudes of the Northern Hemisphere.

The distribution of *Enallagma cyathigerum* may be contrasted with that of the beautiful dragonfly *Sympetrum sanguineum*, sometimes called the ruddy sympetrum (Fig. 1b). This species has a limited distribution in western Europe, parts of Spain, a few places in Asia Minor and North Africa and around the Caspian Sea; it is not found in eastern Asia or North America. In Britain the species is found in only the southern and eastern parts of England, a few places in Wales, and parts of south-eastern Ireland. These areas lie on the northern edge of its distribution, and even here numbers are supplemented by regular immigration of adults from Europe, for the species is a strong flyer. The reason for the limited distribution of this dragonfly is almost certainly that the larva has very precise habitat requirements. The larva is found in ditches and ponds with still waters, but only where certain emergent plants—the great reedmace (*Typha latiflora*) and horsetails (*Equisetum*)—are growing. Why the larva should have these precise requirements is not clear, because it is certainly not a herbivore, and feeds on insect larvae and crustaceans, but so far the larvae have never been found away from the roots of these plants. *S. sanguineum* could therefore be described as a *stenotopic* species—one with very limited ecological tolerance. This inability to colonize any but a very few habitats must certainly have limited its distribution, but other, unknown, factors are also at work, for the species is often absent from waters in which reedmace or horsetails are present.

The northern distribution of these two species may be contrasted with that of the emperor dragonfly, *Anax imperator*[6] (Fig. 1c). The adults of this species are 8–10 cm long, and the larva is found typically in large ponds and lakes and in slow-moving canals and streams; it is a voracious predator and can eat animals as big as fish larvae. Britain lies on the northern edge of the geographical range of this species, where it is found only in the south and east, especially the Norfolk Broads and the Fenland in the east and those southern counties of England where there are many reservoirs and artificial ponds. The distribution covers a band of Europe between about 55°N and 40°N, but unlike the other two species it is well-distributed on the North African coast and the Nile Valley and stretches across Asia Minor to north-west India. It even spreads across the Sahara Desert down into Central Africa, where there are suitable habitats such as Lake Chad and the lakes of East Africa, and it is found in most parts of South Africa except the Kalahari Desert. (It is possible that the South African populations may belong to a separate subspecies from the European forms.) Distributions of this type, occupying two or more areas in widely separated parts of the world, are called *disjunct* distributions.

So the distribution of *A. imperator* is confined to the Old World and does not extend far into Asia. It appears to be basically a Mediterranean and subtropical species whose good powers of flight and fairly broad ecological tolerance have enabled it to cross the unfavourable areas of North Africa to new habitats in southern Africa. No doubt favourable habitats for *A. imperator* do not occur in the other great land masses (although there may be potential competitors there, of course), but the dragonfly cannot now reach them because the land connections lie to the north, where its distribution seems limited.

In most tropical dragonfly species the larvae emerge from the water and metamorphose to the adult at night; they are very vulnerable to predators at the time of emergence and darkness probably affords some protection from birds. But the process of metamorphosis is inhibited by cold temperatures and in northern Europe many species are compelled by low night temperatures to undergo at least part of their emergence in daylight, when birds eat large numbers of them. This probably puts a northern limit to the distribution of many species of dragonfly, including *A. imperator*, which would explain why this species has not been able to invade the Americas or eastern Asia.

The maps in Figure 1 illustrate the contrasts in the distributions of these three species of dragonfly. Even within the British Isles they show very different distributions. This is because all three exhibit, to varying degrees and for a number of different reasons, contrasting preferences for particular habitats. As a result of their geographical position on the north west corner of a great land mass, the British Isles have a complex and variable pattern of climate which interacts with a varied geology to produce a surprisingly wide range of types of habitat for so small an area of the world. Species like these dragonflies reflect this complexity in their own patterns of distribution in Britain.

The cosmopolitan plantain   The plantain, *Plantago major*, has a distribution that could be described as cosmopolitan, because it is found on all the continents except Antarctica[7] (Fig. 2). It is typically a species of grassland habitats, and has a rosette of broad leaves pressed close to the soil surface, from which flower-bearing stems rise. Its

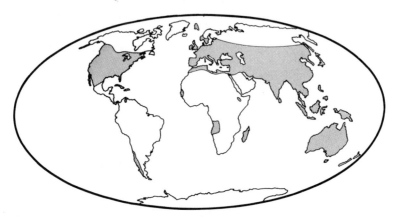

Figure 2. The world distribution of a cosmopolitan species, *Plantago major*.

distribution has not been as thoroughly studied as that of the dragonflies discussed above, but it is found all over Europe except in northern Scandinavia and northern Russia. In Africa it is found along the northern coast, and also in Ethiopia, Angola and Madagascar. It is widespread in North America, being absent from only central Canada and Alaska. The distribution extends across Asia into the south-east and China, and on through the East Indies to Australia and New Zealand; it appears to have spread from here across the Pacific islands to a second area of distribution in the Americas in Chile. On a regional scale the distribution of *P. major* within the British Isles is well-known; the species is found almost everywhere and appears to be quite unaffected by variations in climate or in soil conditions—it is found in grassland on both very acid and very alkaline soils. Clearly this species, like *E. cyathigerum*, is eurytopic. But this alone is not enough to account for its wide distribution; it must have a highly efficient dispersal mechanism. This is provided by its seeds, which are readily eaten by birds and subsequently dropped in a new habitat. The seeds are also resistant to environmental conditions, including those of animals' guts, and at least a proportion of the seeds eaten by animals can germinate after passing through the gut. It is probable that *P. major* is dispersed largely by "hitching a lift" in the gut of an animal that moves about from one habitat to another. It is also likely that man has played a part in the dispersal of this plant, because it is quite likely to be mixed with grasses cut for hay and subsequently carried long distances over land or sea. This seems to be its most likely method of transport across the Pacific Ocean to South America.

Since *P. major* is not in fact found everywhere, what are the factors limiting its distribution? As with many other organisms, the full answer is not yet known. The plant does not extend far into northern regions, but it is found at quite high altitudes elsewhere, and therefore cold is probably not a limiting factor. It is not found in any really arid areas; its broad leaves offer a large surface for evaporation of water and it may become desiccated in dry climates. Local distribution within its grass-land habitat provides some other clues. Figure 3 shows the distribution of the areas in which large numbers of plants are found in a meadow;

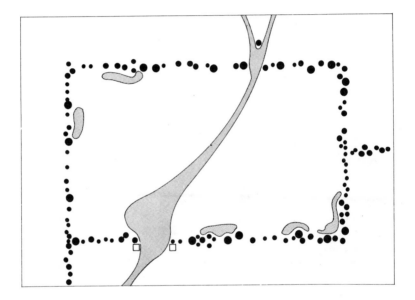

Figure 3. A hypothetical example of the distribution of *Plantago major* in a grazed meadow, shown by shaded areas. Isolated individuals of the plant occur elsewhere in the meadow, but large concentrations of the species occur along paths, around gateways, and in other areas that are trampled and disturbed.

isolated individuals occur elsewhere but cannot be indicated on this scale. In this case the meadow is a hypothetical one but real examples like it could be found nearly anywhere in Britain. It is obvious that the areas of high abundance are around the gateways and footpaths and in a few other areas where grazing animals might collect. This is because the plantain is most common in places where there is intensive grazing by cattle and trampling by these and other animals, including man. It can withstand these pressures better than most grassland plants because of its flattened form. *P. major* probably grows best in such places because it needs full sunlight for photosynthesis in its broad leaves. In other situations it is usually shaded from sunlight by taller plants and is unable to grow well. The species also tends to be an early colonizer of disturbed, bare soil but is eliminated from such places when other species grow tall. The absence of this species from the far northern regions may be explained by the fact that the winter light intensity is too low in those latitudes to enable the plantain to photosynthesize and grow. It is also clear that this species benefits from certain agricultural practices—grazing and ploughing—and it may be that these activities are not sufficiently intensive in some areas of the world, including northern regions, to provide it with enough suitable habitats.

Cycads—living fossils The cycads (family Cycadaceae) are a group of very curious plants; there are 9 genera and about 100 species, all of which are very rare and have a very limited and restricted distribution.[8, 9] Most cycads are trees with a single, short, unbranched stem or trunk crowned with a fringe of fern-like leaves, and in general appearance they are not unlike palm trees (Fig. 4a). A few rather unusual species have the stem buried in the soil with only the top 4 cm and the leaves showing at the surface. There are no flowers, and the seeds are

Fig. 4a

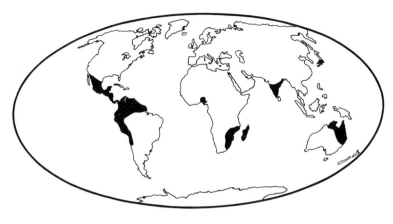

Fig. 4b

Figure 4. (a) A cycad. (b) The world distribution of living species of cycads. The cycads as a whole show a very disjunct or scattered distribution, mainly in tropical regions, and are the relics of a group that was much more widely distributed in the Mesozoic.

carried on structures similar to pine cones; these do not attract insects, and cycads depend on wind for pollination. The entire group is confined to tropical and subtropical regions and, even where they are found, the species are never very common. Cycads are thought to grow very slowly and many plants may be more than 1000 years old. The distributions of individual species are too limited to show on a world map but Figure 4*b* shows the composite distribution of all the species. Four genera are found in North, South and Central America, with two of them confined to Mexico, one species to a small area of western Cuba and another to southern Florida. The largest genus, *Cycas*, has a number of rare species in Australia, India, China, southern Japan, various Pacific islands and Madagascar. Because of their curiosity value, cycads are often grown in ornamental gardens in warm parts of the world, but these planted trees do not, of course, form part of the natural geographical distribution.

How can the rarity and patchy distribution of the cycads be explained? Biogeographers must draw on knowledge from other fields—in this case, from palaeobotany, the study of fossil plants—to understand the distribution of an organism. Palaeobotanists have found many plant fossils similar to the living cycads in rocks that were deposited during the Mesozoic Era, between about 225 million and 65 million years ago. These fossils suggest that the ancestors of modern cycads were not only much more common than their descendants but also much more widely distributed. Fossils of whole groups of cycad species that probably grew together as plant communities—"fossil floras"—have been found in such widely separated places as Oregon, Siberia, Greenland, Sweden, Central Europe, Italy, and Australia, and it is likely that their range was even wider than this list suggests. No such communities of cycad species are found anywhere today; isolated groups of single species are found here and there. Of the places mentioned above where fossil cycads have been found, only Australia now has living representatives.

It seems that cycads were amongst the most important elements of the vegetation in Mesozoic times—the age of the dinosaurs—and it has been suggested that some of the herbivorous dinosaurs browsed extensively on cycads, much as herbivorous mammals feed on flowering plants and trees today. Since the end of the Mesozoic Period the cycads have been reduced to a remnant of their former number of species and extent of distribution. They have been replaced largely by flowering plants (Angiosperms) that probably evolved from fern-like ancestors toward the end of the Palaeozoic Era, but became common only in the late Mesozoic. The cycads had less efficient reproduction and possibly much slower growth than these newer forms and were unable to compete with the flowering plants for space and light.

Since the Mesozoic, tropical climates have been restricted to the equatorial regions and, because the cycads were probably always species of warm environments, this contraction of their habitat must also have been disadvantageous to them. With low rates of reproduction and slow growth, cycads were probably unable to adapt to new conditions fast enough to keep up with climatic changes. The modern species are evolutionary relics whose distribution is limited to parts of the world where conditions are most suitable for them—they can then compete fairly successfully with flowering plants. Many of the areas where cycads survive are isolated geographically, and this may have protected them to some extent against competition from some of the most recently evolved and vigorous groups of flowering plants that have not yet been able to exploit these isolated regions fully.

Climatic relicts Many other species, which in the past were widely distributed, were affected by climatic changes and survive now only in a few "islands" of favourable climate. Such species are called *climatic relicts*—they are not necessarily species with long evolutionary histories, since many major climatic changes have occurred quite recently. The northern hemisphere has an interesting group of *glacial relict* species whose distributions have been modified by the northward retreat of the great ice sheets that extended as far south as the Great Lakes in North America, and to Germany in Europe, during the Pleistocene Ice Ages (the last glaciers retreated from Britain about 10,000 years ago). Many species that were adapted to cold conditions at that time had distributions to the south of the ice sheets almost as far as the Mediterranean. Now that these areas are much warmer, these species survive there only in the coldest places, usually at high altitudes in mountain ranges, and the greater part of their distribution lies far to the north in Scandinavia, Scotland or Iceland. In some cases, species even appear to have become extinct in northern regions and are represented now only by glacial relict populations in the south.

An interesting glacial relict is the springtail, *Tetracanthella arctica* (Insecta, Collembola).[10] This dark-blue insect, only about 1·5 mm long, lives in the surface layers of the soil and in clumps of moss and lichens, where it feeds on dead plant tissues and fungi. It is quite common in the soils of Iceland and Spitzbergen, and has also been found further west in Greenland and in a few places in Arctic Canada. Outside these truly Arctic regions, it is known to occur in only two regions; in the Pyrenean Mountains between France and Spain, and in the Tatra Mountains on the borders of Poland and Czechoslovakia (with isolated finds in the nearby Carpathian Mountains) (Fig. 5). In these mountain ranges the species is found at altitudes of around 2000 metres in Arctic and sub-

Figure 5. The springtail *Tetracanthella arctica*, and a map of its distribution. It is found mostly in northern regions, but populations exist in the Pyrenees and in mountains in central Europe. These populations were isolated at these cold, high altitudes when the ice sheet retreated northwards at the end of the Ice Ages.

Arctic conditions. It is hard to imagine that the species can have colonized these two areas from its main centre further north, because it has very poor powers of distribution (it is quickly killed by low humidity of high temperatures) and is not likely to have been transported there accidentally by man. The likely explanation of the two southern populations is that they are remnants of a much wider distribution in Europe in the Ice Age. But it is surprising that *T. arctica* has not been found at high altitudes in the Alps, despite careful searching by entomologists. Perhaps it has simply not yet been noticed, or perhaps it used to occur there but has since died out. One interesting feature of this species is that whereas animals from the Arctic and the Tatras have eight small eyelets (*ocelli*) on either side of the head, specimens from the Pyrenees have only six. This suggests that the Pyrenean forms have

15

undergone some evolutionary changes since the end of the Ice Ages while they have been isolated from the rest of the species, and perhaps they should be classified as a separate subspecies.

There are probably several hundred species of both animals and plants in Eurasia that are glacial relics of this sort, and they include many species that, in contrast to the springtail, have quite good powers of distribution. One such species is the mountain or varying hare, *Lepus timidus*, a seasonally variable species (its fur is white in the winter and bluish for the rest of the year), which is closely related to the more common brown hare, *Lepus europaeus*. The varying hare has a circumboreal distribution around the northern parts of the world, including Scandinavia, Siberia, northern Japan and northern Canada (although the North American forms are thought by some zoologists to form a separate species, *L. americanus*). The southernmost part of the main distribution is in Scotland and Iceland, but there is a glacial relict population living in the Alps that differs in no important features from those in the more northerly regions. There is, however, an interesting complication—*L. timidus* is found all over Ireland, thriving in a climate that is no colder than that of many parts of continental western Europe. There seems to be no climatic reason why this hare should not have a wider distribution, but probably it is excluded from many areas by its inability to compete with its close relative, the brown hare, for food resources and breeding sites. Relict populations of the varying hare survive in the Alps because, of the two species, it is the better adapted to cold conditions.[11]

The Ice Age closed with a sudden warming of the climate, and the glaciers retreated northwards; behind them came the plant and animal species that had been driven south during glacial times. Warmth-loving animals, particularly insects, were able to move northward rapidly. Plants were slower in their response, because their rate of spread is slower. Seeds were carried northward, germinated, grew, and finally flowered and sent out more seeds to populate the bare northlands. As this migration continued, melting glaciers produced vast quantities of water that poured into the seas, and the ocean levels rose. Some of the early colonizers reached areas by land connections that were later severed by rising sea levels.

The maritime fringe of western Europe must have provided a particularly favourable migration route for southern species during the period following the retreat of the glaciers. Many warmth-loving plants and animals from the Mediterranean region moved northward along this coast and penetrated at least as far as the south-west of Ireland before the English Channel and the Irish Sea had risen to form physical barriers to such movement. The nearness of the sea, together with the influence of the warm Gulf Stream, gives western Ireland a climate that is wet, mild, and frost-free, and this has allowed the survival of certain Mediterranean plants that are scarce or absent in the rest of the British Isles. For example, *Arbutus* (the strawberry tree) is found growing wild nowhere else in Britain, yet it is quite common in western Ireland, especially in County Kerry and County Cork. It also occurs in western Brittany, but its real stronghold is the Mediterranean region, especially Spain and Portugal. Like many Mediterranean trees and shrubs, the strawberry tree is an evergreen and has leathery leaves to reduce water loss in dry summers—which it rarely experiences in Ireland. Another feature showing that *Arbutus* is not native to these northern regions is that it flowers in autumn or early winter. The flowers, which are cream-coloured, conspicuous, and bell-shaped, have nectaries that attract insects, and in Mediterranean areas they are pollinated by long-tongued insects such as bees, which are plentiful in late autumn. In Ireland, however, insects are scarce in the autumn and pollination is therefore much less certain. Thus the strawberry tree reached Ireland soon after the retreat of the glaciers and has since been isolated there as a result of rising oceans. Although the climate has steadily grown colder since its first colonization, *Arbutus* has so far managed to hold its own and survive in this outpost of its range.

Microhabitat Having discussed the reasons underlying the distribution patterns on a world scale and on a country-wide scale, one can finally turn to examine the small-scale pattern of distribution, within a single habitat. It is common to find living together within a habitat several quite closely related species with similar requirements for food and space. If their distribution is examined in detail it is almost always found that each species is living in a distinct microhabitat within the habitat as a whole. As will be seen in Chapter 3, it is probable that this situation has evolved so that the food resources and living space of the habitat can be shared out among its inhabitants without severe competition.

In an area of deciduous woodland in Britain, it is possible, especially in the south, to find as many as twelve or thirteen species of harvest spider (*Order Opiliones*). These are similar in general appearance to true spiders (*Order Araneae*), with eight long, jointed legs, but they have the body fused into a single mass, rather than divided in two by a thin waist as in true spiders. Harvest spiders are 1 cm or less in length, and are voracious predators of any arthropods smaller than themselves, especially fly larvae, springtails, and aphids, and also often of other species of harvest spider and their own young. The distribution of eleven species is shown in Figure 6.

With the exception of *Megabunus diadema*, which is scarce in some places, all these species are quite common in Britain and most have a wide distribution in Europe. For our purposes we can regard the woodland in terms of four rather large divisions—the tree canopy and branches; the shrub layer of small trees up to three metres high (which includes the trunks of taller trees); the herb layer, with plants such as dog's mercury (*Mercurialis*) or bluebells (*Endymion*); and the ground layer, containing leaf litter, mosses, stones, dead wood, and so on. Each of these divisions or horizontal layers in fact contains many micro-

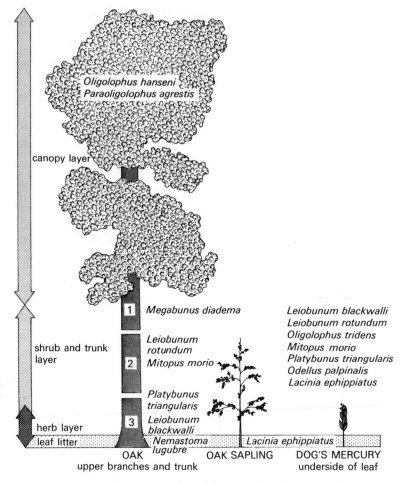

Figure 6. A model of the distribution of eleven species of harvest spider (order Opiliones) in woodland in southern England. Each species is most common in one or more microhabitats within the habitat but is sometimes to be found elsewhere.

habitats, but harvest spiders are quite large and active animals and their distribution cannot be considered on a smaller scale. Each species has one or two levels at which it is found much more frequently than elsewhere, as shown in Figure 6.

Many more species (at least seven) occur frequently in the herb layer than elsewhere. This is probably because this layer is more varied in structure than the others and contains more small-scale microhabitats. Three of these species are rather different from the others in their behaviour. The immature individuals of *Leiobunum rotundum* spend the spring in the herb layer, but at about the end of June they move to tree trunks, where they spend the day about 10 metres above ground level. At night, however, they leave the tree trunks and return to the herb layer to hunt, so the species has separate microhabitats for nesting and for feeding. *L. blackwalli* is also found on tree trunks as well as in the herb layer, but is not known to migrate like *L. rotundum*, and tends to occur rather lower on the trunks. A third species, *Mitopus morio*, is found in all microhabitats except the soil and the higher parts of the canopy, and this wide distribution is probably related to the fact that it is largely a predator of other harvest spiders. Not all seven species are likely to be present and competing for resources at the same time—their periods of activity are rather spread out through the year. The species showing perhaps the strongest habitat preference is *Nemastoma lugubre*, which is found only in leaf litter and humus. The English ecologist, Dr Valerie Todd, made an interesting study of the ecology of harvest-spiders and, in particular, of their reactions to the relative humidity (water content) of the air in their habitats. In a careful experimental study, Dr Todd offered the various species a choice of environments with different atmospheric humidities.[12] She found that those species which live in very humid microhabitats, such as *Nemastoma lugubre* which lives in the litter layer, or *Oligolophus tridens*, which lives in the ground vegetation, tended to select the most humid environments. Species like *Leiobunum rotundum* and *L. blackwalli*, which live mostly in less humid places above the ground vegetation, selected drier environments. The species normally found in drier places, up on the branches of trees, such as *Oligolophus hanseni* and *Paraoligolophus agrestis*, selected the most dry environments. These experiments suggest that the harvest spiders are adapted to the atmosphere of the microhabitats where they are most common, and that they suffer physiological stress elsewhere, by dehydration in drier places and by wetting in moister places. They are therefore limited to their microhabitats by their own physiology; this problem will be considered in more detail in Chapter 2.

The distribution of harvest spiders, with each species having one or more microhabitats as a "headquarters", but sometimes occurring elsewhere in the habitat, is probably typical of most groups of animals in most habitats. Probably, the same is true for plant species, such as *Plantago major*, already discussed, which grows best and is most common on open, trampled areas of a meadow although isolated individuals may occur in other places. Most habitats contain plant species, such as mosses, which are adapted to live in their wetter or more shaded parts. However, the degree to which species are adapted and restricted to particular habitats or microhabitats varies greatly amongst both plant and animal species. A few general tendencies can be recognized. For instance, plant species such as trees, that are big and slow-growing and which themselves exert a great influence on the form and structure of the habitat, tend to show less strict habitat or microhabitat preferences than small and delicate species such as herbs and mosses. Amongst animals, there are broad differences between predators and herbivores. Predatory animals are usually active, relatively large species and are often not very specific in their food requirements, so they often tend to have rather broad microhabitat preferences. Herbivores, on the other hand, are less mobile and often more specific in their food requirements and, of the species in a habitat, these are usually the ones with the most precise microhabitat preferences.

Distribution and food   The distribution of an animal may be very closely related to that of the food it eats. A particularly interesting example is that of the Chalkhill blue butterfly, *Lysandra coridon*, which is on the wing in July, August, and September.[13] In Britain it is found only in relatively few areas in southern England, although it is widespread in mainland Europe. When the distribution is analysed it is found to coincide closely with areas of countryside underlain with chalk in central southern England and the Hampshire Downs, the North and South Downs and the Chiltern Hills of south-eastern England, and the Cotswold Hills in the west Midlands (Fig. 7). Why should the range of *L. coridon* be related to the geology of the land in this way? The most significant factor in determining this pattern is that the caterpillar larva of this butterfly usually feeds on only one plant,

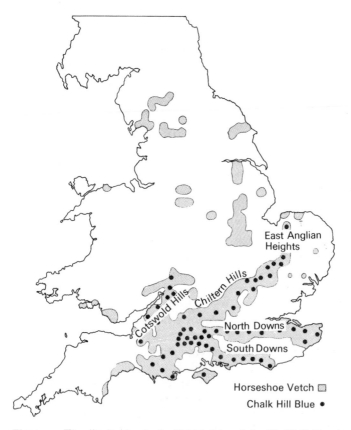

Figure 7. The distribution in the British Isles of the Chalkhill Blue butterfly, *Lysandra coridon* (each dot representing an area where the species is known to breed), and of its foodplant, the horseshoe vetch, *Hippocrepis comosa*.

Hippocrepis comosa, the horseshoe vetch, which is itself rather uncommon. This plant is one of a group called *calcicoles*, which are able to grow well only on chalky or limestone-rich soils. The vetch is a perennial plant that develops a mat of foliage up to 1 metre or so across and 30 cm high. The caterpillars of the Chalkhill blue feed most actively on the leaves of the plant during May and June. When fully developed they drop to the soil, crawl into cracks, and metamorphose into pupae. After one or two weeks the adults emerge from the pupae, crawl up the vegetation and, after drying their wings, take flight. The females lay eggs on the leaves of the vetch, which hatch into caterpillars in the following April, and these live on the plant throughout the summer, growing rapidly: thus the life cycle repeats itself.

The reason why calcicoles such as the horseshoe vetch grow only on chalky soils has been one of the most intractable problems of ecology, and is still not yet fully understood. Chalky soils contain large amounts of calcium compounds derived from the underlying chalk or limestone, and these combine with the carbonic acid (a combination of water with carbon dioxide) contained in rain, and with acids derived from the chemical breakdown of plant tissues in the soil, and produce a very alkaline environment. Some other compounds—especially phosphates and nitrates, which are essential to plants—are in relatively short supply in chalky soils, and it seems that plants need special adaptations to absorb them from the soil under these very alkaline conditions. Plants adapted in this way have an advantage over others in the competition for these scarce nutrients, and may be the only ones that can grow on these soils. It seems that the horseshoe vetch is one of these especially adapted plants—experiments show that its roots grow much faster in chalky soils than in others. This would explain the limited distribution of the vetch and hence of its dependent herbivore, the butterfly. But other problems remain, because the distribution of the butterfly is not as wide as that of its food plant. Not only is the Chalkhill blue absent from many places in the south where the vetch occurs, but it is not present (except very occasionally) in any of its food plant's more northern habitats on limestone. In the south the butterfly appears to be more common on south-facing slopes of the Downs, and it may be that it can live only in these warmer habitats that get plenty of sun. Since the Chalkhill blue is more common in Europe, it is probably a Mediterranean species that in the southern parts of the British Isles is at the northern limit of its distribution. The Chalkhill blue also has a curious relationship with some species of ant, especially the yellow meadow ant, *Lasius flavus*. These ants drink a sugary fluid produced by glands on the bodies of the caterpillars, and in return they protect them from insect predators and spiders, and sometimes even carry them to more favourable locations of their food plants. This ant is known to be able to build its nests only in certain situations and on certain soils. Thus it may be that because of their peculiar relationship, the distribution of the Chalkhill blue is limited by that of the ant, as well as by the distribution of the vetch.

There remains the problem of why the butterfly should be limited to a single food plant. One might expect this to be a disadvantage, because, as in this case, the distributions of many plant species are limited by geology and soil, and by climatic factors. Ecologists think that in most cases, animals feeding on a single food plant are not limited to it by special nutritional requirements that it alone provides. More probably, such animals gain a competitive advantage over others by being especially adapted to exploit a single food source with maximum efficiency, even though this has disadvantages if the distribution of the plant is limited. The host plant may provide favourable physical conditions, and many insects have also developed special camouflages, effective only against the background provided by one food plant. Also, herbivores probably "recognize" their food plants by the presence of particular chemical compounds, usually not themselves of nutritional importance. These "token stimuli" are especially important to species limited to a single food plant, but what these factors are in the case of the Chalkhill blue is not known.

Influence of man In contrast to the small worlds of individual species of harvest spider or butterfly, the effects of man are global— he is the most important species in the ecology of most of the world's surface. Human numbers have increased steadily over the last 10,000 years, but the rate of increase has been much greater in the last 400 years, so that there are now roughly 3250 million people in the world. This success is partly due to the fact that man is a very eurytopic species— he can live in a large number of different habitats, but is able to do this mostly because of his great technical skills, which allow him to modify the conditions of the environment to suit his needs. Human manipulation of the environment—fire, agriculture, industry, and urbanization— has increased as dramatically as human numbers, and all this activity has influenced and modified the geographical distribution of thousands of species of animals and plants. The history of the effect of man on each other species is unique, but we can recognize a few broad relationships.[14]

First there are the plants and animals that man has deliberately taken with him around the world because he uses them as suppliers of raw materials for food, transport, or hunting. There are several hundred of these "domesticated" species. Many have not only been distributed around the world but have also, accidentally or by deliberate selective breeding, been changed in their characteristics so that they can live in many different environments. A good example is the potato, *Solanum tuberosum*, which was originally cultivated by the Indians on the high plateaux of the Andes in South America. It was brought to Europe by the Spaniards in about 1550, and eventually spread back across the Atlantic to North America. The potato needs a cool climate with fairly high rainfall, but varieties have been developed that can grow in colder and wetter, or in warmer and drier regions. The species is now grown in all temperate regions, and as a food species is second in importance only to rice. Many domestic animals have a similar history; the domestic fowl, *Gallus domesticus*, was originally domesticated during pre-Christian times in north-west India and Malaya, and now includes many different varieties that have been taken everywhere by man. Like many other domesticated species the fowl seems to have been kept first of all for use in religious ceremonies and for cock-fighting; only later were its egg-laying and meat-producing qualities recognized. Similarly, rice (*Oryza sativa*) may have begun its association with man as a weed in the wetter parts of other cereal crops. This group also includes species that, although not part of man's economy, have been introduced to new areas as pets or for other, largely sentimental, reasons—an example is the house sparrow (or English sparrow), *Passer domesticus*, introduced by European settlers to North America, Australia, and New Zealand.

"Exotic" animals and plants have often been introduced into Europe and North America by early naturalists, returning explorers, and rich land-owners—for example, several ornamental pheasants and the rhododendron (*Rhododendron ponticum*) introduced into Britain. Many of these animals have escaped and established themselves more or less firmly in their new homes—wild populations of the Canada goose, *Branta canadensis*, are now a familiar sight over much of Britain and the grey squirrel, *Sciurus carolinensis*, first introduced to Britain from North America in 1876, has now become a serious pest to the forestry industry.

A second group of species is that which has been accidentally spread by man. Many of these are important because, although they may be quite harmless in their original environments, they have become serious pest species in their new ones. The fluted or cottony cushion scale insect, *Icerya purchasi*, comes originally from Australia, where it feeds—apparently causing little damage—on citrus fruit trees. In the late 19th century it began to appear in other places where citrus fruit was grown, presumably having been transported on trees or machinery. Populations developed in California and the southern United States, South America, Hawaii, Japan, New Zealand, India, South and North Africa, and southern Europe. In all these places it has, at various times, become a serious pest, threatening the whole citrus industry. Many other species of pest, both animals and plants, have similar histories.

Study of the biology of the cottony cushion scale insect in Australia showed that it did not become a pest there because its numbers were kept low through predation by several insects, including a ladybird beetle, *Rodolia cardinalis*. This beetle has subsequently been introduced into all the areas where scale insects were causing damage and has reduced their numbers to levels at which they are no longer an important pest. *Rodolia cardinalis* is an example of the third group—predatory or parasitic species that man has taken from one part of the world to another to control the numbers of pest species; this is known as *biological control*. The job would otherwise be done by using pesticides—at far greater cost and with the risk of pollution. Hitherto, biological control methods have been little used (when compared with the massive and often disastrous application of pesticides) but it is likely that this new form of domestication will become more and more important.

A further group of species is of those that have been made much rarer or extinct as a result of man's activities. Some of these, such as the dodo, the great auk, and the grizzly bear, stood no chance against man as a predator. Others, such as the flightless rails of the Pacific islands, have become extinct through failure to resist predators or to compete successfully with animals introduced by man, usually cats and rats. Many of these were ecologically unsuccessful species with very limited distributions, which would probably have become extinct sooner or later in any case.

This leads to what is probably the biggest group of species whose distributions have been affected by man, those whose habitats have disappeared or been modified as a result of agricultural or industrial practices (including pesticides and pollution) and urbanization. There have been a few studies of this process, but some of the more dramatic examples are well known—such as the disappearance of the swallow-tail and large copper butterflies from the British Isles as a result of the drainage of fenland in East Anglia. Sometimes animals have been able to use the artificial environment created around them by man, and have increased in number and sometimes in range. An excellent example is that of the black redstart, *Phoenicurus ochruros*, a Mediterranean species (originally a cliff-nesting bird) that has been able to extend its range to Britain mainly because of its ability to colonize city buildings, particularly bomb-sites, thereby taking advantage of a habitat created by man.

The history of man's increasing influence over the fortunes of the fauna and flora with which he shares the world will be discussed in more detail in chapters 8 and 9, which will also outline the changes in his attitudes and activities which man must make if he is to prevent irrevocable destructive changes from taking place in the forseeable future.

References

1 NEAL E. (1948) *The Badger*. Penguin Books, Harmondsworth.
2 STRESEMANN E. & NOWAK E. (1958) Die Ausbreitung der Turkentaube in Asien und Europa. *J. Ornithol.* **99,** 243–296.
3 *Book of British Birds* (1969) Reader's Digest Association, London.
4 MAYR E. (1965) The nature of colonization in birds. In: Baker H.G. and Stebbins G.L. (eds.). *The Genetics of Colonizing Species.* Academic Press, New York and London.
5 CORBET P.S., LONGFIELD C. & MOORE N.W. (1960) *Dragonflies.* New Naturalist Series. Collins, London.
6 CORBET P.S. (1957) The life history of the Emperor Dragonfly, *Anax imperator* Leach (Odonata : Aeshnidae). *J. Anim. Ecol.* **26,** 1–69.
7 SAGAR G.R. & HARPER J.L. (1964) Biological flora of the British Isles. *Plantago major* L., *P. media* L. and *P. lanceolata* L. *J. Ecol.* **52,** 189–221.
8 CHAMBERLAIN C.J. (1919) *The Living Cycads.* University of Chicago Press, Chicago.

9   FOSTER A.S. & GIFFORD E.M. (1959) *Comparative Morphology of Vascular Plants*. W.H. Freeman & Co, San Francisco.

10   CASSAGNAU P. (1959) Contribution à la connaissance du genre *Tetracanthella* Schott. *Mém. Mus. nat. Hist. natur., Paris (Zool.)* **16** (7), 201–260.

11   SOUTHERN H.N., ed. (1964) *The Handbook of British Mammals*. Blackwell Scientific Publications, Oxford.

12   TODD V. (1949) The habits and ecology of the British harvestmen (Arachnida, Opiliones) with special reference to those of the Oxford district. *J. Anim. Ecol.* **18,** 209–216.

13   FORD E.B. (1945) *Butterflies*. New Naturalist Series, Collins, London.

14   ELTON C.S. (1958) *The Ecology of Invasions*. Methuen. London.

CHAPTER 2

THE PHYSICAL LIMITATIONS OF LIFE

In Chapter 1 environmental conditions were often referred to as determining or limiting the distribution of plants and animals. It was suggested that these *limiting factors* in the environment include *physical factors* such as temperature, light, wetness, and dryness, as well as *biotic factors* such as competition, predation, or the presence or absence of suitable food. In this and in the following chapter the ways in which such factors influence organisms will be described in more detail.

First, though, the meaning of the term "limiting factor" must be understood. Anything that tends to make it more difficult for a species to live, grow, or reproduce in its environment is a limiting factor for the species in that environment. To be limiting, such a factor need not necessarily be lethal for a species; it may simply make the working of a species' physiology or behaviour less efficient, so that it is less able to reproduce or to compete with other species for food or living space. For instance, we suggested in Chapter 1 that a northern limit may be set to the distribution of certain dragonflies by low night-time temperatures. In the more southerly parts of northern regions, at least, temperatures are not so low that they kill dragonflies *directly*, but they are low enough at night to force the insects to metamorphose during the day, when they are more vulnerable to predatory birds. In this case, then, the limiting factor of temperature does not operate directly but is connected with a biotic environmental factor, that of predation. Many other limiting factors act in a similar way.

Environmental gradients Many physical and biotic factors affect any species of organism, but each can be considered as forming a *gradient*. For example, the physical factor of temperature affects species over a range from low temperatures at one extreme to high temperatures at the other, and this constitutes a temperature gradient. These gradients exist in all environments and affect all the species in each environment. As seen in Chapter 1, different species vary in their tolerance of environmental factors, being either *eurytopic* (ecologically tolerant) or *stenotopic* (ecologically intolerant), but each species can function efficiently over only a more or less limited part of each gradient. Within this *range of optimum*, the species can survive and maintain a large population; beyond it, toward both the low and the high end of the gradient, the species suffers increasing physiological stress—it may stay alive, but because it cannot function efficiently, it can maintain only low populations. These areas of the gradient are bordered by the upper and lower

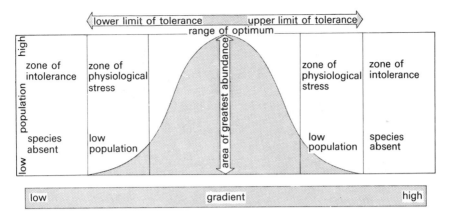

Figure 8. Graphical model of the population abundance maintained by a species of animal or plant along a gradient of a physical factor in its environment. (From Kendeigh, S. C., 1961. Animal Ecology, p. 11. Prentice Hall, Englewood Cliffs, New Jersey.)

limits of tolerance of the species to the environmental factor. Beyond these limits, the species cannot survive because conditions are too extreme; individuals may live there for short periods but will either die or pass quickly through to a more favourable area (Fig. 8).

A relatively simple example of an environmental gradient is the global gradient in temperature which runs from hot equatorial regions northward to cold Arctic areas. The gradient has a great many local variations due to local climatic conditions, but in general there is a progression from hot to cold in average temperature through the year. The animals and plants adapted to live in cool temperate conditions can obviously survive only in those parts of the global temperature gradient where such conditions are found. These cool-temperate regions, then, lie in the species' range of optimum, to the north of which lie areas that are too cold, and to the south of which lie regions that are too hot. In the southernmost parts of the cold end of the gradient, cool-temperate species may maintain low populations especially in favourable years, but further north these populations peter out as conditions become too cold. There will be similar areas with low populations in the northernmost regions of the warm end of the gradient.

The English ecologist, Sir Edward Salisbury, made some careful studies in the 1920's of the distribution of some plants common in western Europe and the British Isles, and compared these distributions with those of various climatic factors.[1] One species he studied was the wild madder, *Rubia peregrina,* a creeping woody plant growing in scrubby country, hedgerows and the like. In Europe this species has a predominately southern and western distribution, and is most abundant in Spain and Italy; it is also fairly common in parts of west and central France. The fringes of the species' distribution to the north are the extreme south and southwest of the British Isles (including Ireland),

Figure 9. The distribution of the wild madder, *Rubia peregrina*, and of the January Isotherm 4.5 C.

the foothills of the Alps in northern Italy and the Adriatic coast of Yugoslavia. To the south its limits are the Mediterranean coast of North Africa as far east as Tunisia. Along the fringes of its distribution it is a distinctly uncommon or even rare plant. In the British Isles, for instance, where its distribution is well known, it occurs in only 21 of the 112 vice-counties (roughly equal divisions of the country for the purpose of recording plant distributions) and in Ireland in 10 out of 40, and it is never common even where found. Salisbury found that throughout Europe the northern boundary of the distribution of *R. peregrina* coincided closely with a line joining all those places where the long-term average air temperature in January was about 4·5°c (the January 4·5°c *isotherm*) (Fig. 9). He suggested that this was not mere coincidence but that temperatures in January were critical for this species since this is the period when new shoots for the following spring are formed, and this is just the kind of complex biochemical and physiological process very likely to be inhibited by low temperatures. This January 4·5°c isotherm thus marks the limit of tolerance of *Rubia peregrina* to low temperatures, and the species is not found beyond it to the north or east. The extreme south and west of Britain is a zone of physiological stress for the plant, where it can maintain only low populations. In years when January temperatures in this area are appreciably lower than average, *R.peregrina* will probably be unable to form new shoots or able to do so only much later than usual, thereby shortening the growing and flowering season. Further south, in western France and Spain, lies the range of optimum of this species where high populations are maintained and good growth is possible every year. This is a neat example because the limits of distribution of the organism correspond so closely with those of a physical factor, but Salisbury's hypothesis lacks experimental proof and, furthermore, it is not known if temperature factors limit the southern distribution of this temperate species in North Africa.

Although temperature is one of the most important because of its effect on the metabolic rate of organisms, many other physical factors in the environment are limiting ones. A whole family of factors is related to the amount of water present in the environment. Aquatic organisms obviously require water as the basic medium of their existence, but most terrestrial animals and plants, too, are limited by the wetness or dryness of the habitat, and often also by the humidity of the atmosphere, which in turn affects its "drying power" (or more precisely, the rate of evaporation of water from the ground and from animals and plants). Light is of fundamental importance because it provides the energy that green plants fix into carbohydrates during photosynthesis, thus obtaining energy for themselves (and ultimately for all other organisms). But light in its daily and seasonal fluctuation also regulates the activities of many animals. The concentrations of oxygen and carbon dioxide in the water or air surrounding organisms are also important. Oxygen is essential to most animals and plants for the release of energy from food by respiration, and carbon dioxide is vital because it is used as the raw material in the photosynthesis of carbohydrates by plants. Pressure is important to aquatic organisms; deep-sea animals are specially adapted to live at high pressures, but the tissues of species living in more shallow waters would be easily damaged by such pressures.

In marine environments, variation in the salinity of the water affects many organisms, because many marine organisms have body fluids with much the same salt concentration as sea-water (about 35 parts per thousand), in which their body tissues are adapted to function efficiently. If they become immersed in a less saline medium (in estuaries, for instance), water moves into their tissues due to the physical process called *osmosis*, by which water passes from a dilute solution of a salt to a concentrated one. If the organisms cannot control the passage of water into their bodies, the body fluids are flooded and their tissues can no longer function. This problem of salinity is an important factor in preventing marine organisms from invading rivers, or freshwater ones from invading the sea and spreading across oceans to other continents (Fig. 10).

Interaction of factors The environment of any species consists of an extremely complicated series of interacting gradients of all the

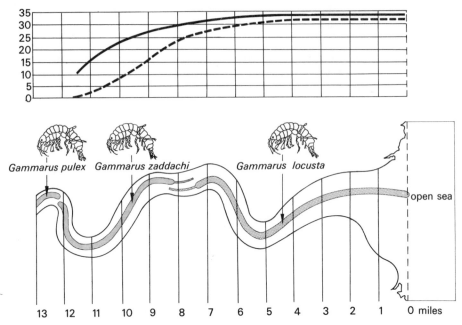

Figure 10. The distribution along a river of three closely-related species of amphipod (Crustacea), relative to the concentration of salt in the water. *Gammarus locusta* is an estuarine species and is found in regions where salt concentration does not fall below about 25 parts per thousand. *G. zaddachi* is a species with a moderate tolerance of salt water and is found along a stretch of water between 8 and 12 miles from the river mouth where salt concentrations average 10–20 parts per thousand. *G. pulex* is a true freshwater species and does not occur at all in parts of the river showing any influence of the tide or salt water.[2]

factors, biotic as well as physical, and these influence its distribution and abundance. Populations of the species can live only in those areas where favourable parts of the environmental gradients that affect it overlap. Factors that fall outside this favourable region are limiting ones for the species in that environment.

Some of the interactions between the various factors in an organism's environment may be very complex and difficult for the ecologist to interpret or for the experimentalist to investigate. This is because a series of interacting factors may have more extreme effects on the behaviour and physiology of a species than any factor alone. To take a simple example, temperature and water interact strongly on organisms, because both high and low temperatures reduce the amount of water in an environment, high temperatures causing evaporation and low ones causing freezing, but it may be very hard to discover if an organism is being affected by heat or cold or by water lack. Similarly, light energy in the form of sunlight exerts a great influence on organisms because of its importance in photosynthesis and in vision, but it also has a heating effect on the atmosphere and on surfaces, and therefore raises temperatures. In natural situations it is often almost impossible to tell which of many possible limiting factors is mainly responsible for the distribution of a particular species.

An interesting example of the complexity of interaction between environmental factors was studied by the American ecologist Dr Michael R. Warburg, in his work on two species of woodlice (sowbugs or slaters, Crustacea, Isopoda) living in rather dry habitats in southern Arizona.[3, 4] One species, *Armadillidium vulgare*, is found mostly in grasslands and scrubby woodland and is also widely distributed in similar habitats elsewhere in North America and in Europe. The other, *Venezilla arizonicus*, is a rather rare species, confined to south-eastern U.S.A. and found in very arid country, with stony soil and a sparse vegetation of cactus and acacia. Warburg investigated the reactions of these two species to three environmental factors, temperature, atmospheric humidity and light. His experimental techniques involved the use of a simple apparatus, the choice chamber or *"preferendum"* apparatus, in which animals may be placed in a controlled gradient of an environmental factor. The behaviour of the animals, particularly the direction in which they move and their speed, can then be used to suggest which part of the gradient they find most satisfactory; this is termed the *preferendum* of these particular animals in this particular gradient. Warburg's method for testing the interactions of light, temperature and humidity on the woodlice was the classic scientific approach of isolating the effects of each factor separately and then testing them two or three at a time in all possible combinations. For instance, he might set up a gradient of temperature between hot and cold and test the reactions of the animals to this either with the whole gradient at a low humidity (dry) or the whole gradient at a high humidity (wet), or with the hot end of the gradient dry and the cold wet, or with the cold end dry and the hot end wet. He might then test the effect of light on these four situations by exposing each in turn to constant illumination, constant darkness, or one end of the gradient in darkness and the other in light. Such work is extremely time-consuming and requires great patience.

Warburg found that in general *A. vulgare* prefers low temperatures (around 10–15°C), high humidities (above 70 per cent relative humidity, i.e. the air is 70 per cent saturated with water vapour) and is rather weakly attracted to light. This accords well with what is known of the species' habitat and habits—it lives in fairly humid, cool places and is active during the day. *Venezilla arizonicus*, on the other hand, prefers lower humidities (around 45 per cent), higher temperatures (20–25°C) and will generally move away from the light. Again, this accords well with the species' habits, since it lives in rather dry, warm places and is active at night. The reactions of the species change, however, and become harder to interpret when they are exposed to more extreme conditions. For instance, at high temperatures (35–40°C), *A. vulgare* tends to choose lower humidities, irrespective of whether these were in light or dark. One of several possible explanations for this behaviour is that at these high temperatures the species' physiological processes can only be maintained if body temperatures are lowered by permitting loss of water vapour from the body surface, which is more rapid at lower humidities. The normal reaction of *V. arizonicus* changes, if the species is exposed to very high humidities; it then tends to move to dryer conditions even if these are in the light. Warburg concludes that for these two species light is not really an important physiological factor and acts mostly as a "token stimulus", a clue to where optimum conditions of humidity and temperature may be found. For *V. arizonicus*, which lives in a dry or *xeric* habitat, darkness indicates the likely presence of the high temperatures and low humidities it prefers. For *A. vulgare*, in its cooler, more humid or *mesic* habitat, there is little risk of desiccation except in the most exposed situations and the species can afford to be relatively indifferent to light.

Warburg's study indicates the great complexity of the reactions of even relatively simple invertebrate animals such as Crustacea to the physical factors of their environment. If we analyse also the biotic factors of the animal's environment, food and enemies, the picture becomes even more complex. Other studies of *Armadillidium vulgare*, for instance, in California, indicate that the species shows quite strong preferences for different types of food (mostly various types of dead vegetation) and these also influence its distribution.[5]

Plant species, too, show strong reactions to the physical factors of their environment but, because of their inability, for the most part, to move in response to them, they react by changes in their pattern of germination or growth. Seeds normally require rather precise conditions of temperature, soil moisture and light for successful germination. Adult plants respond to the environment in the size and shape to which they grow, and changes in their environmental conditions produce faster growth or periods of dormancy. Interaction of factors is also important in plants. Green plants require sunlight for their photosynthesis and, up to certain levels, photosynthesis increases with increase in light intensity. But light also has a heating effect on the tissues of the plant and, since many of the chemical reactions within the tissues operate well only at fairly low temperatures, very high light intensities tend to inhibit photosynthesis or even to damage the plant.

Patterns of climate Many of the most important physical factors of the land environment have very distinct patterns of variation in different parts of the world. This pattern we call *climate*. The climate of an area is the whole range of weather conditions, temperature, rainfall, evaporation, water, sunlight, and wind that it experiences through all the seaons of the year. Many factors are involved in the determination of the climate of an area, particularly latitude, altitude, and position in relation to seas and land-masses. The climate in turn largely determines the species of plants and animals that can live in an area.[6]

Climate varies with latitude for two reasons. The first reason is that the spherical form of the earth results in an uneven distribution of solar energy with respect to latitude, as shown in Figure 11a. As the angle of

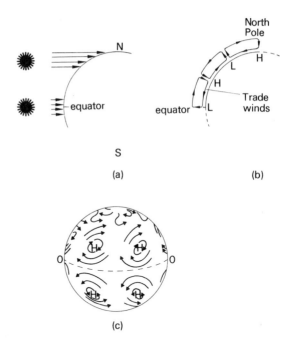

Figure 11. Patterns of climate. (a) Due to the spherical shape of the earth, polar regions receive less solar energy per unit area than the equatorial regions. (b) The major patterns of circulating air masses (cells) in the Northern Hemisphere. (c) The general pattern of air movements on the surface of the earth, showing the Coriolis effect. (H = high pressure, L = low pressure)

incidence of the sun's rays approaches 90°, the area over which the energy is spread is reduced, so that there is an increased heating effect. In the high latitudes, energy is spread over a wide area; thus polar climates are cold. The precise latitude that receives sunlight at 90° at noon varies during the year; it is at the Equator during March and September, at the Tropic of Cancer (23·45°N) during June, and at the Tropic of Capricorn (23·45°S) during December. The effect of this seasonal fluctuation is more profound in some regions than in others.

The second reason is that variations also result from the pattern of movement of air masses. Figure 11b shows an idealized picture that assumes a uniform surface to the Earth. Under these conditions air is heated over the equator, and therefore rises (causing a low pressure area) and moves northwards. As it moves northwards it gradually cools and increases in density until it descends, where it forms a sub-tropical region of high density. Air from this high pressure area either moves towards the equator, forming the Trade Winds, or else moves polewards. This latter air eventually meets cold air currents moving south from the polar region, over which air is cooled and descending (causing a high pressure area). Where these two air masses meet, a region of unstable low pressure results, in which the weather is changeable.

This idealized picture is complicated by the *Coriolis effect* (named in honour of the French mathematician Gaspard Coriolis, 1792–1843, who analysed it), which results from the east–west rotation of the Earth. This force tends to deflect a moving object to the right of its course in the Northern Hemisphere and to the left in the Southern Hemisphere. Thus the horizontal component of the movement of air masses in our ideal picture can be represented as in Figure 11c.

The distribution of oceans and land-masses modifies this simple picture yet further. Because heat is gained or released more slowly by water than by land-masses, heat exchange is slower in maritime regions, while at the same time humidities are higher. In summer, therefore, continental areas tend to develop low-pressure systems as a result of the heating of land-masses and the conduction of this heat to the overlying air masses. Conversely, in winter the reverse situation occurs, continental areas becoming cold faster than the oceans, and high pressure systems developing over them. Because most of the Earth's land areas occur in the Northern Hemisphere, the ideal situation shown in the diagram is disrupted to a far greater extent in the Northern than in the Southern Hemisphere.

In addition to the heating and cooling effects of land-masses, climate is also affected by altitude. On average the air temperature falls by 0·6°C for every 100-metre rise in height, but this varies considerably according to prevailing conditions, especially the aspect and steepness of slope and the wind exposure. Because of this tendency for temperature to fall with increasing altitude, the organisms inhabiting high tropical mountains—such as Mount Kenya in East Africa—may be more like the flora and fauna of colder regions than that of the surrounding lowlands[7] (Fig. 12). However, although in general temperature falls as one ascends such mountains, other environmental conditions do not mirror precisely those found at higher latitudes. For example, the seasonal variations in day-length typical of high latitude tundra areas

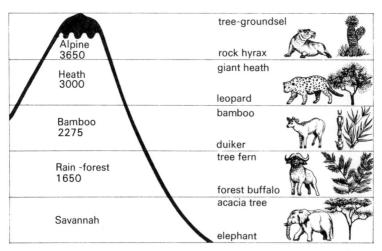

Figure 12. Altitudinal zonation of plants and animals on Mount Kenya. This illustrates how within a given latitude the biome which develops will be dependent upon such factors as altitude.

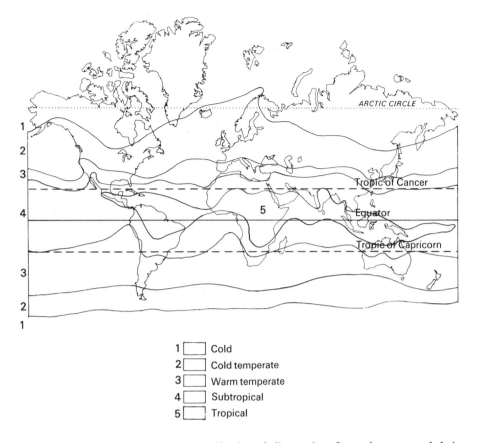

1 ☐ Cold
2 ☐ Cold temperate
3 ☐ Warm temperate
4 ☐ Subtropical
5 ☐ Tropical

Figure 13. A simple classification of climates into five major types and their distribution on the surface of the earth.

are not found in the "alpine" regions of tropical mountains. Also the high degree of insolation resulting from the high angle of the sun produces considerable diurnal fluctuations in temperature which are not found in tundra regions. It is not surprising therefore that the altitudinal zonation of plants and animals should not reflect precisely the global, latitudinal zonation.

The climate of an area, then, is the result of the many varying factors that affect the region, and the Earth's surface accordingly experiences a great variety of climates distributed over it in an intricate pattern; they can be classified into five divisions, though each contains scores of regional variants. These are shown in Figure 13.

Biomes and life-forms Each of these climatic types and their major sub-divisions has a number of characteristic plant and animal communities that have evolved so that they are well-adapted to the range of environmental factors in them; such characteristic communities are called *biomes*. The distinctions between biomes are not

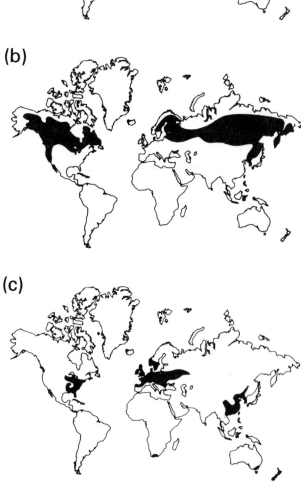

(a)

(b)

Figure 14. Distribution of the major terrestrial biomes of the world. (a) Tundra. (b) Northern coniferous forest. (c) Temperate forest. (d) Tropical rain forest (including monsoon forest). (e) Temperate grasslands (light shading) and tropical savannah (dark shading). (f) Desert. For further explanation see text.

(c)

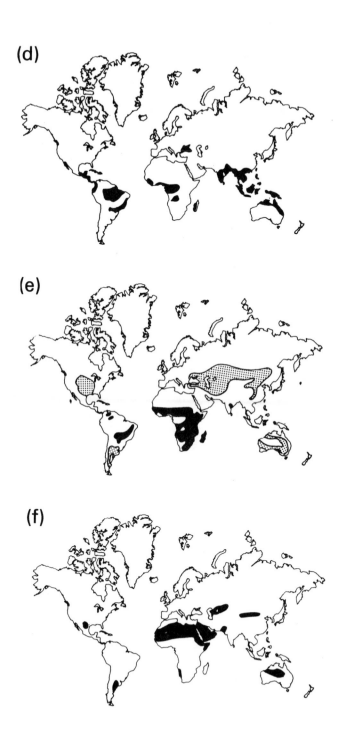

(d)

(e)

(f)

necessarily related to the taxonomic classification of the organisms they contain, but rather to the *life-form* (the form, structure, habits, and type of life-history of the organism in response to its environment) of their plants and animals. This concept of the life-form was first put forward by the Danish botanist Christen Raunkaier[8] in 1903. He observed that the most common or dominant types of plants in a climatic region had a form well suited to survive in prevailing conditions. Thus in Arctic conditions, the most common plants are dwarf shrubs and other low-growing plants; these have no extensive above-ground growth that would be broken by heavy winter snowfalls, and their buds are carried at or just below the surface of the soil where they obtain the maximum protection from cold and wind in the long winter. In warmer climates, the characteristic types of vegetation are trees or tall shrubs that carry their buds and reproductive structures well above the ground because they are rarely exposed to severe weather conditions. Deserts usually contain small plants, mostly quick-growing annuals, with little above-ground growth, and buds and survival structures below the soil surface, because of the risk of drought. Animals also show distinct life-forms adapted to different climates, with cold-resistant, seasonal, or hibernating forms in cold regions and forms with drought-resistant skins or cuticles in deserts. Nevertheless, animal life-forms are usually far less easy to recognize than are those of plants and, consequently, most biomes are distinguished by the plants they contain and are named after their dominant life-form.

There is no real agreement among biogeographers about the number of biomes in the world. This is because it is often difficult to tell whether a particular type of vegetation is really a distinct form or is merely an early stage of development of another, and also because many types of vegetation have been much modified by the activities of man.

We shall describe eight terrestrial biomes, a freshwater one, and a marine one[9] (Fig. 14).

Tundra (Fig. 15) is found around the Arctic Circle, north of the tree-line. Smaller areas occur in the Southern Hemisphere on sub-Antarctic islands. Alpine tundra occurs above the tree-line on high mountains, including those in the tropics. It is the most continuous of biomes and the easiest to define. Winter temperatures are $-57°C$ or lower: water melts at the soil surface in summer (air temperature is rarely over $15°C$) but there is always a permanent layer of frozen soil underneath—the *permafrost*. There is a very short growing season, and only cold-tolerant plants can survive. Typical plants are mosses, lichens, sedges, and dwarf trees. Large herbivores include reindeer, caribou, and musk ox. Small herbivores include snow-shoe hares, lemmings, and voles. Many birds migrate there from the south in summer, feeding on the large insect populations in the tundra during that season. Carnivores are Arctic fox, wolves, hawks, falcons, and owls.

Figure 15. Tundra with caribou, interior of Alaska, Mt. McKinley Peak, August.

Northern Coniferous Forest (Taiga) (Fig. 16) forms an almost unbroken belt across the whole of northern North America and Eurasia—and is one of the most extensive biomes. Its northern margin with the tundra is sharp, being the Arctic tree-line, but its southern limit is less definite —taiga is also found on high mountains in lower latitudes, such as the southern Rockies. In the northern forests, the winters are long and cold, the summers short and often very warm. The soil in winter is mostly frozen to a depth of about 2 metres, but thick snow cover can keep soil temperatures as high as $-7°$c. Trees are mostly evergreen conifers, able to photosynthesize all year and to resist drought (a result of strong winds and extreme cold) with their needle-shaped waxy leaves. They remain undamaged by snowfalls because of their overall shape. Taiga usually contains vast tracts of one or two tree species only, except along

39

Figure 16. Northern coniferous forest area; Canadian Rocky Mountain.

rivers. The soil is *podsol* (see page 50) and invariably contains a layer of ash-white sand, due to the leaching-out of bases and clays by humic acids (organic acids produced by the decay of plant material). Animals in this biome are limited by severe winters and the small number of different habitats. The most important large herbivores are deer—more species live here than in any other biome. Rodents are plentiful and can burrow under snow and survive harsh winters. Carnivores include wolves, lynxes, wolverines, weasels, mink, and sable; omnivorous bears are also found. Birds either are adapted to feeding in taiga, such as crossbill, or are summer migrants feeding on the vast seasonal swarms of insects.

40

Figure 17. Temperate forest; Herbst in Odenwald.

Temperate forest (Fig. 17)   Vast tracts of the taiga are still in the natural climax state, but little climax forest remains in the temperate forest biome. There are 4 basic types of temperate forest. (1) Mixed forest of conifers and broad-leaf deciduous trees. This was the original climax vegetation of much of north-central Europe, eastern Asia, and north-east North America—little remains today. (2) Mixed forests of conifers and broad-leaf evergreens. This once covered much of the Mediterranean lands but very little is left. It still occurs in the Southern Hemisphere, in Chile, New Zealand, Tasmania, and South Africa. (3) Broad-leaf forests almost entirely of deciduous trees. This formerly covered much of Europe, northern Asia, and eastern North America, and is found in the Southern Hemisphere only in Patagonia. (4) The rare broad-leaf forest consisting almost entirely of evergreens. This occurs throughout much of Florida, and also in north-east Mexico and in Japan. In the Southern Hemisphere it occurs on the southern tip of South Island, New Zealand. All these regions have very high rainfall, and the dripping forests have been termed "temperate rain-forests". In all temperate forests, there is frequently an understorey of saplings, shrubs, and tall herbs, which is particularly well-developed near the forest edge or where human interference has occurred. Temperate forests have warm summers but cold winters, except on western sea-boards. Winter temperatures may fall below freezing-point. The deciduous trees escape these cold winters by losing their leaves; many plants have underground over-wintering organs. The fauna includes bears, wild boar, badgers, squirrels, woodchucks, many insectivores, and rodents. Predators include wolves and wild cats (on the decline), red foxes, and owls. Large herbivores are the deer. This biome is extremely rich in bird species, especially woodpeckers, titmice, thrushes, warblers, and finches.

Tropical rain-forest (Fig. 18) occurs between the Tropics of Cancer and Capricorn in areas where temperatures and light intensity are always high and rainfall is greater than 200 cm a year (and is at least 12 cm in the driest month). Because of this, there is a great variety of trees: in some parts of the Brazilian rain-forests, there are as many as 300 species of trees in 2 sq. km. The popular image of the jungle—thick, steamy, and impenetrable—is borne out only in those areas that man has at some time cleared, especially along river margins; true climax tropical forest has very little undergrowth. The canopy is extremely dense; the light intensity below may be as low as 1 per cent of that above, and thus only a few extremely shade-tolerant plants can survive there. Life is concentrated in the canopy, where there is plenty of light. The crowns of the trees are covered with *epiphytes*—plants that use the trees only for support and are not parasites. *Lianas*—vines rooted in the ground but with leaves and flowers in the canopy—are also characteristic. Dead plants are rapidly decomposed, so there is little undecayed plant matter on the forest floor. The rate of turnover of nutrients is very high and the tropical forest has a higher productivity than that of any other terrestrial biome. The tropical rain forest biome contains the greatest variety of animal life of any biome, because of the richness of the food resources that it offers and the relative constancy of the conditions of the environment through the year. There is a great profusion of birds with many different diets—seeds, fruit, buds, nectar or insects. Many of the mammals are adapted to arboreal life (monkeys, sloths, ant-eaters, many small carnivores) but there are also many ground-living forms, including rodents, deer and peccaries. Amphibia, and reptiles, especially snakes, are important as predators of small vertebrates and invertebrates.

Temperate grassland occurs in regions where rainfall is intermediate between those of desert and of temperate forest, and where there is a

Figure 18. Tropical forest biome; Puerto Rico.

fairly long dry season. Temperate grassland has many local names—the *prairies* of North America, the *steppes* of Eurasia, the *pampas* of South America, and the *veld* of South Africa—but the dominant plants in all of them are the grasses, the most widespread and successful group of land plants. The soil always contains a thick layer of humus, unlike forest soils, but is more exposed than the latter, and therefore more likely to dry out. The dominant animals are large grazing mammals—on the North American prairies, vast herds of bison and prong-horn (which man had virtually wiped out by the close of the last century, but is now reintroducing); over the steppes of Eurasia, the saiga antelope, wild horse, and wild ass once roamed in herds; in the South American pampas, the natural grazer is the guanaco; and in Australia, the kangaroos fill this role. All these have been largely replaced by man with

43

Figure 19. Savannah biome; Kenya.

domestic grazing animals, often with disastrous results, as we shall see
in Chapter 8, although grasses are adapted to withstand the effects of
natural grazing.

Tropical grassland or savannah (Fig. 19) is a term applied to any
tropical vegetation ranging from pure grassland to woodland with
much grass. It covers a wide belt on either side of the Equator between
the Tropics of Cancer and Capricorn. The climate is always very warm
and there is a long dry season, and thus the plants often have drought-
resisting features. The grass is much longer than that of temperate
grassland, growing to $3\frac{1}{2}$ metres. There is often a great variety of trees,
which also show drought-resisting features; a typical group is the
acacias. The dominant animals are large grazing mammals, the African

savannah having the greatest variety, and burrowing rodents are also found. Large carnivores, such as lions and hyenas, prey on the grazers.

Chaparral occurs where there are mild wet winters and pronounced summer droughts (known as Mediterranean climate), and in areas with less rain than grasslands. The vegetation is *sclerophyllous* (hard-leaf) scrub of low-growing woody plants, mainly evergreen, with hard, thick, waxy leaves—adaptations to drought. In the Northern Hemisphere it occurs mainly in countries fringing the Mediterranean basin, but also in north-west Mexico and California. Formerly this biome had a varied flora and fauna, with many herbivores such as ground squirrels, deer, and elk, and mountain lions and wolves as their predators, but this has been greatly reduced by man. In the Southern Hemisphere there are small areas of chaparral in southern Australia, southern Chile, and South Africa.

Deserts (Fig. 20) are areas experiencing extreme drought. A good definition is those areas where rainfall is less than 25 cm per year, or—if higher—is mostly lost immediately by evaporation. Deserts can be divided into hot deserts (such as the Sahara) with very high daytime temperatures, often over 50°C, and low night-time temperatures below 20°C with relatively mild winters, and cold deserts (such as the Gobi Desert in Mongolia) with very severe winters and long periods of extreme cold. Typical desert has large areas of barren rock or sand and very sparse vegetation. Desert plants are adapted to drought in various ways: some have drought-resistant seeds; others have small thick leaves that are shed in dry periods; yet others, such as the New World cacti, are succulents, storing water in their stems. Desert animals are mostly small enough to hide under stones or in burrows during the intense daytime heat in hot deserts. Certain rodents are well adapted to desert life—they live in cool burrows, are largely nocturnal, and waste very little water in their urine. Insects and reptiles lose little water, having waterproof skins and excreting almost dry, crystalline urine. Deserts spread when wind carries the top sand away, or when man encourages his domestic animals to overgraze their edges.

Freshwater biomes are far less self-contained than those of the surrounding land or the open sea. They receive a continual supply of nutrients from the land, but much of this is washed downstream in the rivers and there is an overall loss of organic material. Thus they are generally far less rich in nutrients than oceans, and usually less productive than either sea or land environments. They are more changeable than ocean or land biomes; rivers gradually wear away the land through which they pass and thus the river biome itself gradually changes, and many small ponds are seasonal, drying up in summer. There is a wide range of freshwater environments ranging from small ponds and streams to vast lakes and wide rivers. At the lower end of the scale, they are often better considered merely as a wet extension of the surrounding terrestrial biome.

Figure 20. Desert biome; Algeria.

The dominant plants of larger lakes and slow rivers are phytoplankton, but larger floating and rooted plants cover considerable areas. Many of the animals are restricted to the freshwater habitat; amphibians, though living on land, need fresh water in which to breed; land animals use fresh water for drinking and bathing; and many birds are adapted to the freshwater habitats. Animal communities in large lakes correspond to planktonic, nektonic, and benthic communities of the oceanic biome (see below). Some large lakes have well-defined shores, constituting sub-biomes: examples are the Great Lakes with their dune systems, or Lake Victoria with its muddy shores. Marshes (salt marshes and freshwater marshes) are best considered as intermediate between marine or freshwater biomes and the surrounding terrestrial biome, and estuaries are transitional both between freshwater and marine biomes and also between the water and the land. They have a very complex structure and are highly productive, with a great variety of plant and animal life. Freshwater habitats have suffered greatly from pollution by man—toxic industrial wastes, detergents, and vast quantities of sewage are dumped into rivers and lakes, and cause the extinction of all but a few resistant forms of life.

Marine biomes Land covers only 29 per cent of the earth's surface, whereas the oceans take up 71 per cent, with an average depth of 3900 metres. It is impossible to distinguish regional biomes in the seas, because of the uniformity of the marine environment and of the ease of distribution of its inhabitants. On the land, animals and plants of different latitudes have different life-forms; in the sea, animals do have distinctive forms, but these vary according to the depth at which they live, rather than according to latitude—for example, deep-sea animals are of a life-form especially adapted to cope with high pressures and total darkness. Water has a higher specific heat than soil or rock, and even the warmest oceans never reach the high temperatures of tropical forests or hot deserts. Similarly, the coldest seas are never as cold as the tundra or northern forests. The surface temperature is never greater than 30°c and rarely falls below 0°c. Marine organisms obviously have no problems in obtaining sufficient water, but light is a limiting factor. The tiny photosynthetic plants (*phytoplankton*) are restricted to the upper photic zone (the uppermost 200 metres); virtually no light penetrates below 500 metres. Atmospheric oxygen and carbon dioxide are plentiful at the surface and these gases are also dissolved in the water. Pressure is an important factor limiting the downward movement of shallow-water species. Sea-water is much richer in nutrients than fresh water, and these are recycled to the photic zone by upwellings of deep currents. In some other areas, surface waters converge and descend. Where these are already exhausted of nutrients, the area of descent forms a "desert", such as the Sargasso Sea in the southern North Atlantic. Such areas are the only virtually unproductive parts of the surface waters.

There are three principal marine biomes. (1) The *oceanic biome* of open water, away from the immediate influence of the shore. This is further divided into the *planktonic sub-biome* containing free-floating plankton (mostly microscopic organisms with buoyancy mechanisms); the *nektonic sub-biome* of active swimmers, including fish, squids, turtles, and marine mammals; and the *benthic sub-biome*, whose fauna is especially adapted for life on the sea floor. (2) The *rocky shore biome*, dominated by large algae that show zonation up the shore (see Fig. 21). Life here is in constant danger of desiccation when uncovered by the water. (3) The *muddy or sandy shore biome* where mud and sand is constantly being washed ashore by the sea, and the animals are often in danger of being buried. The main plants are thin green algae growing in flat sheets on the shore, such as the sea-lettuce (*Ulva*). Animals include burrowing worms, and also wading birds, which are important predators of the invertebrate fauna.

Figure 21. Zonation of seaweed; South Devon, England, seashore—(Top) *Pelvetia canaliculata* (Channelled wrack). (Middle) *Fucus spiralis* (Flat wrack). (Bottom) *Ascophyllum nodosum* (Knotted wrack).

Soil The distribution of soils is another important aspect of biogeography. Terrestrial plants are mostly rooted in the soil and obtain from it water and their nutrients, such as nitrates and phosphates. Because all animals are ultimately dependent on plants for food, they are also in turn dependent on the soil. Many factors influence the ability of plants to root and take up nutrients from the soil. Most important are the structure of the soil and its texture (the size of the particles of which it is composed), the amount of nutrients actually present, and the quantity of water with gases dissolved in it, and air spaces, that it contains. These are different in different types of soil. It is not surprising, therefore, that soil has a strong influence on the distribution of living things. But the relationship between organisms and soils is a complex one, because the soil is largely produced by interaction between the organisms themselves and climate.

The soil profile can be divided into a series of horizons which are usually given the letters A, B and C (Fig. 22). That closest to the surface, the A horizon, is one of intense biological activity; it can be subdivided into horizons A_0, A_1 and A_2. The superficial A_0 horizon

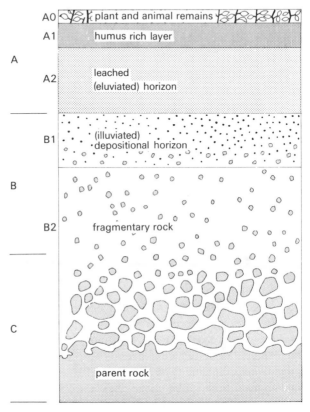

Figure 22. A generalized diagram of a soil profile. Under conditions of leaching, A and B horizons become apparent as a result of the removal of materials such as iron and aluminium from the A_2 horizon and their deposition in the B_1 horizon. This process is termed podsolization.

consists of dead plant and animal tissues which are undergoing decomposition by bacteria and fungi as well as being broken up and partially digested by earthworms and other soil animals. These animals mix the organic matter with weathered mineral particles to form a humus-rich A_1 layer. The A_2 and B layers become evident when a soil is subjected to a high degree of "leaching"—percolation of rain water which is slightly acidic due to the presence of carbonic and organic acids. The degree of percolation is a function of such factors as the porosity of the soil (large particle soils drain more freely), the climate (a high precipitation : evaporation ratio will result in increased leaching) and the surface vegetation (this may control the acidity of the surface litter layer).

Under conditions of high leaching the process known as *podsolization* occurs. This involves the mobilization of organic colloids in the A_2 horizon together with the breakdown of the soil clay fraction and the removal of iron and aluminium from the clay particles. As a result the A_1 horizon becomes rich in residual silica and bleached because of iron removal. Iron, aluminium and organic matter may be re-deposited lower in the profile, in the B_2 horizon. This deposition may result in the formation of a hard concretion or "iron pan", which may cause impeded drainage and water-logging of the A horizons. Brown earth soils, where A and B horizons cannot be distinguished, occur largely under deciduous woodland, whereas the podsol type of profile is associated with coniferous woodland and heath vegetation.

Whenever a new habitat is created—whether by volcanic action, mountain building, or melting of ice sheets—the surface exposed is generally one of rock. The action of climate on the rock surface—heating by the sun, cooling and dissolving of salts by rain, or shattering by frost—causes it to "craze" or break up so that many different sized crevices are formed. The shelter from wind, sun, and rain that these provide enables small plants, especially algae and mosses to colonize the rock surface, and the food resources and additional shelter that these provide are soon exploited by various small animals, particularly protozoans, rotifers, and small insects. As the organisms go through their life histories and die, they leave behind them organic remains (*humus*) that mix with the broken rock to produce an increasingly fertile soil for new generations of plants and animals. Eventually the food resources of the habitat are adequate for invasion by larger plants such as grasses and small herbs, which can put roots deep into the crevices in the rock to obtain nutrients and water, causing further break-up of the rock. As more and more plant species enter the community, they provide an increasing variety and quantity of food resources, so that many animal species follow them. In time, shrubs and perhaps trees and all their associated animals colonize the habitat, which ultimately develops the flora and fauna characteristic of the biome of the climatic region in which it is situated.

Ecological succession This process of evolution of the biotic community and its habitat is called *ecological succession,* and is itself an important factor in the distribution of organisms; it will be considered in more detail in the next chapter. Throughout this process of ecological succession, the soil is developing along with the living things it supports, with mineral particles and inorganic nutrients derived from the parent rock and organic matter derived from its biotic community. During succession the soil becomes deeper and more complex in structure, until it reaches a form characteristic of its biome. This form is determined partly by the type of vegetation of the biome and partly by the amount of nutrients that that vegetation draws from the soil and the amount of organic matter that it returns. But climate is also a determining factor. Soils in areas with a high rainfall are often water-logged, so that the concentration of oxygen in the soil is too low, and the concentration of carbon dioxide too high, for satisfactory growth of plant roots, which in turn limits the amount of organic matter that the plants can add. In better-drained areas with high rainfall, leaching may remove many of the nutrients in the soil, either along the surface or deeper into the underlying rock, and this greatly reduces the soil's fertility. In hot regions, evaporation of water from the soil surface may keep it too dry to permit plants to root. These climatic factors also directly influence the biotic community—for instance, by determining whether earthworms can survive. These animals, which play an important role in mixing humus and mineral material, are rarely able to live in soils that are too dry, too acid, or badly aerated.

The nature of the rock from which the mineral material has been derived also affects the type of soil that is formed. Rocks that are very porous to water, such as chalk or limestone, develop well-drained, often dry, soils, whereas more impervious rocks produce badly drained marshy soils. The acidity of the rock is also important; limestone and chalk produce very alkaline soils, but other rocks, such as granite, tend to produce acid soils. The size of the mineral particles in the soil, which influences the soil's water-holding properties and the efficiency with which plants can absorb nutrients from it, is also influenced by the nature of its parent rock. Reactions between each soil and the climate tend to produce a characteristic soil in each biome.

Not *all* the climatic region occupied by a particular biome is covered with the vegetation that is typical of the biome, nor does it necessarily have the typical soil. For instance, the map (Fig. 14c) shows the eastern and south-eastern United States, and western Europe, as part of the Temperate Deciduous Forest biome, but much of this area is actually occupied by grassland or scrub vegetation. One reason for this is that much of the forest has been cleared by man (mostly since European colonization in the case of North America), and the soil ploughed and used for agriculture. If farmland is abandoned (as in large areas of Georgia and the Carolinas), it follows over some 200 years an ecological succession back to the biotic community characteristic of its biome—in this case, temperate deciduous forest. The community present at the end of succession in a biome is called the *climax community* of the biome, and ones earlier in ecological succession are called *seral stages*. All biomes contain communities in all seral stages from bare rock to the climax community, and the whole sequence of communities involved, the *sere*, is just as characteristic of the biome as is the climax. (Processes of change and succession probably continue after the *climax* community has been reached, but these are generally much slower than earlier ones.)

Microclimates Therefore climate, through its effects on organisms and on the development of soils, imposes an overall pattern of distribution of living things on a global and regional scale. Within the biome or climatic region, however, organisms are far from evenly distributed, either between different habitats or within them. One reason for this is the great variation on this scale in the distribution of climate, soil, and other physical factors. The drainage of soil or the amount of wind to which it is exposed may determine the type of vegetation that grows on it, or at least the rate at which the community may evolve toward the climax typical of that region. Small-scale variation in surface features of the environment may also affect community development. South-facing slopes in the Northern Hemisphere receive more sun than north-facing ones, for instance, and thus are warmer and tend to be drier. They are also often more sheltered from cold north-east winds, and these factors influence the biotic community that develops on them. In grassland, for instance, insects, spiders, and some plants are frequently much more abundant on such south-facing slopes.

The vegetation itself greatly influences the climate of the community. Tall plants, for instance, provide shade and protection from wind and rain for smaller members of the community. Areas beneath trees or in the lee of hedgerows generally have a less variable, warmer and more humid climate than more exposed places. Areas of natural vegetation, such as forests or grasslands, have a distinctive climate, which is a modification of the general climate typical of the geographical region in which they are situated. These modifications are the result both of small-scale geographical variation in terrain—slope, aspect, etc.—and of shading and deflection of wind and rain by the community itself. Such local climates are called *microclimates*.[11]

Important factors in the development of microclimates are the extent of penetration into the vegetation of sunlight energy in the form of heat, and of wind and rain (which cause cooling and control humidity). This can be illustrated by examining the climatic conditions that might be found in a grassland sward including a dense layer of clover, at noon on a fine summer day, perhaps after early morning rain. Figure 23 shows the conditions of light, temperature, wind speed and relative humidity that might be present at various levels in the vegetation. Bright sunlight does not penetrate far into the vegetation: halfway down in the grass the intensity of light is only about half that above the vegetation. The amount of light falls off even more sharply once the layer of clover is reached: above the clover, the light is at about 35 per cent of its full strength; below, it falls to only a few per cent, and at the soil surface there is almost total darkness. Wind speed is also much reduced by the vegetation, and beneath the clover there is almost no air movement at all. This factor particularly influences the relative humidity of the air. Above the vegetation the water evaporated from the soil and the plants is blown away, so the air is only about half saturated with water vapour. Where the vegetation is more dense, below about 15 cm and especially amongst the clover, water vapour is trapped and the humidity is much

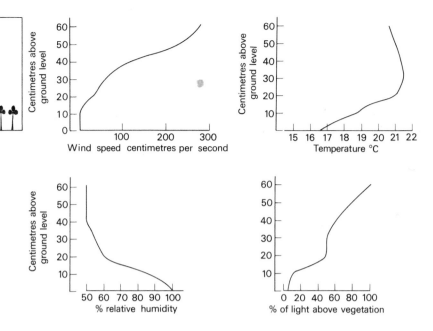

Figure 23. Diagram showing the structure of grassland vegetation and the effect which this has upon the microclimate of the habitat.

higher, and at soil level the air is fully saturated with water vapour. The distribution of temperature at various levels in the vegetation is in fact rather more complicated than this, because it is a result of inter-action between the heating effect of sunlight, cooling by wind, and cooling by the damp surface of the soil. The warmest place is about half way up the vegetation, where sunlight and wind speed are both about half strength. Above this, conditions are cooler because the wind is stronger; below, conditions are much colder, as well as more humid.

Variations of microclimate, such as these within vegetation, have a considerable effect on the distribution of small animals. The activities of herbivorous insects such as caterpillars, grasshoppers, and froghoppers are very much influenced by the conditions of humidity and tempera-ture among the grasses on which they feed. If conditions become too hot or too dry, they may become over-heated or desiccated, and must retreat to cooler, moister regions below, where there may be little suitable food. Cold and rain may also drive the insects down to the base of the vegetation. The amount of light present in the vegetation is important for such predatory animals as spiders, which must be able to see their prey. Some animals, such as most species of woodlice, can survive only in conditions of very high air humidity, and most of the time are confined to the soil surface; only in very damp conditions—just after rain, for example—can they move up into the vegetation.

The microclimate of grassland is very much influenced by the height and density of the vegetation. When the height of the vegetation varies

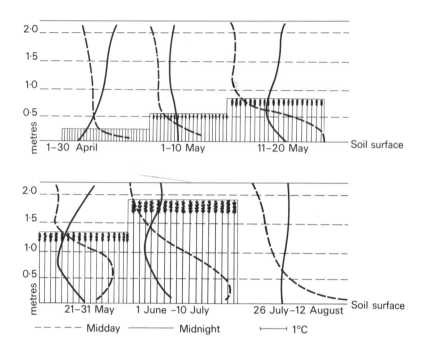

Figure 24. The temperature profile of a field of winter rye at midday and at midnight at different periods of its growth. The vertical profile lines show the temperature at different heights in and above the crop, the temperature increasing as the line moves towards the right. For instance in April, when the crop is only about 0·25 m high, the warmest level at midday is the still air just above the soil surface, and the temperature is much lower in the moving air above the crop; at night, as the ground cools, the situation is reversed. The midday temperature difference between the cool air above the crop and the warm air within it increases until, by late May, the crop has become so thick that sunlight does not penetrate to the soil surface. The 0·5 m level is therefore the warmest at midday, but is also the coolest at midnight when lower levels are warmed by heat radiated from the ground. By June and early July, though the crop is now higher, the sun is now nearly overhead and its heat therefore penetrates further. The warmest midday level is now only 0·25 m above the soil and, since the soil now gets warmer by day, its night-time radiation of heat is higher, so that the coolest midnight level is higher. In late July, when the crop has been harvested, the temperature levels are typical of a bare field: the soil, exposed to the sun, is the warmest midday level, but night temperatures are almost the same at all levels.

(Adapted from Geiger, *The Climate Near the Ground.*)

in different seasons, the microhabitats will also vary. Figure 24 shows how temperatures vary at different levels by day and night in a crop of rye (which is rather like a simple grassland) during its development from small seedlings in April to the harvesting of the crop in July.

The effect of vegetation on microclimates is at its most extreme in woodland or forest. Here the vegetation may be anything up to 30 metres or more high, and the tree-top canopy is often the densest part of the vegetation. The canopy reflects or absorbs much of the sunlight reaching the habitat, and is often almost impenetrable to wind. Figure

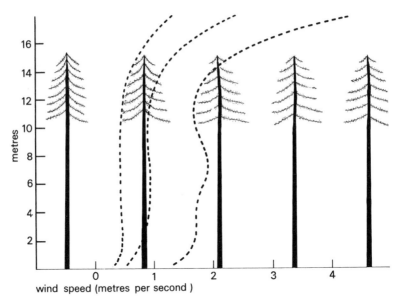

Figure 25. Wind profiles in a pine forest at three different wind speeds. (After Geiger, 1965.)

25 shows the wind speeds at different levels within a pine forest at different wind speeds above the canopy. It can be seen that the canopy has the effect of reducing air movements, and in this way maintains uniform microclimatic conditions. If the wind speed above the canopy is very high, the air at tree-trunk level starts to move, though at a lower rate. Within woodland in summer, conditions tend to be cool, because sunlight penetrates so little, and it is also damp, because the lack of wind keeps humidity high. There is also much less fluctuation in conditions through the day than in more open habitats, and this effect is strongest at the forest floor. Woodland microclimates favour organisms that prefer low light intensities and humid air, and that are intolerant of the greater fluctuation in conditions found in other habitats. The distributions of the woodland harvest spiders (discussed in Chapter 1) are strongly influenced by humidity. All require the humid conditions of woodland, but within woodland the various species have preferences for different levels of humidity. In winter in temperate forest, when leaves are shed, the canopy disappears and light and wind can penetrate closer to the forest floor, so that differences between woodland microclimates and those of other habitats may be less extreme (Fig. 26).

By the process of ecological succession, which has already been discussed, plant and animal communities change or develop from simple ones with short vegetation in the early stages, to climax communities such as deciduous forest. During this development the microclimate of the habitat tends to become progressively more complex and more different from that of the surrounding environment, as the plants of the community get larger and more complex in structure. As the

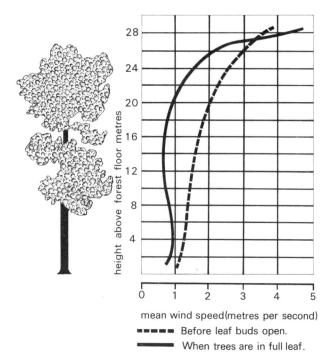

mean wind speed(metres per second)

▬ ▬ ▬ ▬ Before leaf buds open.

▬▬▬▬▬ When trees are in full leaf.

Figure 26. These graphs of wind speed at different levels in an oak forest show that the wind is more able to penetrate into lower levels of the forest before the leaves open. (After Geiger, 1965.)

microclimate changes, new species, better adapted to the new conditions than the existing species, are able to invade the community. These new species may compete so successfully with the earlier ones, that they may replace them, and the species composition of the community changes. In this way the community controls not only its microclimate but also the species of which it is composed. This self-regulating mechanism is a fundamental way in which biological factors affect the distribution of species; more of these factors will be considered in the following chapter.

References

1 SALISBURY E.J. (1926) The geographical distribution of plants in relation to climatic factors. *Geogr. J.* **67**, 312–335.
2 SPOONER G.M. (1947) The distribution of *Gammarus* species in estuaries, Part I. *J. Mar. Biol. Ass.* **27**, 1–52.
3 WARBURG M.R. (1964) The responses of isopods towards temperature, humidity and light. *Animal Behaviour* **12**, 175–186.
4 WARBURG M.R. (1968) Behavioural adaptations of terrestrial isopods. *Am. Zoologist* **8**, 545–559.
5 PARIS O.H. & SIKORA A. (1967) Radiotracer analysis of the trophic dynamics of natural isopod populations. In: Petensewica K. (ed.) *Secondary Productivity of Terrestrial Ecosystems*, Volume II. Warsaw.

6 LOWRY W.P. (1967) *Weather and Life: An Introduction to Biometeorology.* Academic Press, New York and London.
7 EYRE S.R. (1968) *Vegetation and Soils: a World Picture,* 2nd edn. Edward Arnold, London.
8 RAUNKAIER C. (1934) *The Life Forms of Plants and Statistical Plant Geography.* Clarendon Press, Oxford.
9 ODUM E.P. (1971) *Fundamentals of Ecology,* 3rd edn. W.B. Saunders & Co, London and Toronto.
10 MONEY D.C. (1965) *Climate, Soils and Vegetation.* University Tutorial Press, London.
11 GEIGER R. (1965) *The Climate Near the Ground.* Harvard University Press, Cambridge, Mass.

CHAPTER 3

MAKING A LIVING

In the last chapter the influence of physical factors in the environment on the distribution of animal and plant species was considered and it was found that these formed barriers, of varying degrees of effectiveness, at both large and small scales. In this chapter, some of the ways in which a species' relationship with other species (including its food) may create similar *biological* barriers, will be examined. To understand the nature of these, the biogeographer must turn to the work of ecologists, who have made many studies, both in the field and the laboratory, of the interaction of organisms, both within and between species.

Food and feeding   One of the most important biological barriers is the availability of food. As mentioned in Chapter 2, plants require from their environment adequate supplies of water, inorganic nutrients and energy in the form of sunlight. From these they build organic compounds by the process of photosynthesis, and the availability of these "food materials" determines the distribution of many photosynthetic species. The amount of solar energy available may be especially important; species of plant adapted to live in open country cannot invade woodland habitats because shading by trees reduces the available sunlight to levels at which they cannot photosynthesize efficiently.

Many animal species have limited distributions that are associated with the distribution of the plants or other animals on which they feed. Although the caterpillars of the Chalkhill blue butterfly, *Lysandra coridon* (whose distribution was discussed in Chapter 1) may feed in the laboratory on the leaves of one or two species of plants, in the field they are found exclusively on the vetch, *Hippocrepis comosa*—and it was noted that the limited distribution of this plant is a major factor in the distribution of the butterfly. Species like this with very narrow food preferences (*stenophagous* species) are much more likely to be limited in their distribution by their food than are species with very broad food preferences (*euryphagous* species). But species with such catholic tastes as the American bobwhite quail (*Colinus virginianus*) which was found to feed on the fruits and seeds of 927 different plant species (as well as on many insects and spiders) are unlikely to have distributions limited by food preferences. In general herbivorous species have more restricted food requirements, and therefore distributions more limited by food, than do carnivores. One reason for this might be that in most habitats there is a much smaller quantity of animal food available to the carnivorous species than plant food for herbivores, and if carnivores had very limited food preferences they might often have to go for long periods without food. But even among herbivorous species there is great variation in the degree of restriction to particular food. For instance, in a survey of the feeding habits of 240 species of aphids or "plant-lice" which suck plant-juices, it was found that 27 per cent of them were restricted to a single species of plant, 40 per cent to a group of closely-related species and 33 per cent fed on many different species. Amongst herbivores as a whole the most common types are probably those that feed on a group of closely related species.

A good example of a herbivore of this type is the Colorado potato beetle, *Leptinotarsa decemlineata*, which is found naturally in the eastern Rocky Mountain region where it feeds mostly on the wild sand-bur (*Solanum rostratum*). When a closely-related species, the domesticated potato (*Solanum tuberosum*) was widely introduced from its original home in the Andes, *L. decemlineata* was able to use this new food supply, and has thereby been able to extend its distribution right across North America and even to many parts of Europe (Fig. 27).

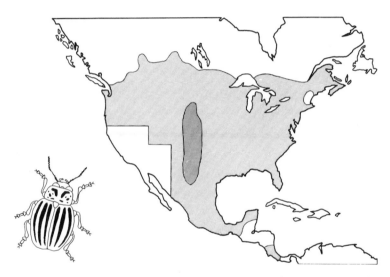

Figure 27. Distribution of the Colorado beetle in North and Central America (original distribution in dark shading, 1962 distribution in light shading).

There is, however, an exception to the rule that carnivores tend to be less limited by food than herbivores; this is to be found in those animal species that are parasites of others. Most of these can obtain their food and the living conditions that they require from only one, or at best very few species of host. Their distributions, therefore, are strongly limited by that of their hosts.

Competition Another biological factor influencing or limiting the distribution of a species is competition with other species for food and space. Although there is a great variety of different food materials available in most habitats, the total quantity of food available is strictly limited. Even our own species, with all its technological and agricultural skill, is currently discovering how limited are the available resources of food, energy and living space. Many other animal species are short of food at certain times—during unfavourable seasons, or periods when their numbers are too high in relation to the food supply. The abundance or shortage of food influences not only the numbers of individuals of a particular species in a habitat, but also the number of species that can co-exist there. The ability of a species to extend its geographical range, or its local range into a new habitat, depends on its ability to find unexploited food supplies, or alternatively, to compete successfully with, and partially or completely displace another species. Animals compete most frequently for food but also sometimes for territorial space, nest-building sites or burrows. In plants competition occurs mainly for space, both in the soil for root systems and above it for access to sunlight.

But it is no easy matter to prove that competition is a major factor limiting the distribution of species. This is because, in the wild, the effects of competition between species are hard to observe; they occur over long periods of time, and changes in the numbers of two or more species which might be due to competition are often masked by changes caused by weather or predators. Occasionally, we can observe the effects on other species when a new species invades their habitat naturally or is deliberately introduced by man. Sometimes the new species is so successful in competition for resources with an established species that it may eliminate it. This effect is called "*competitive displacement*". At other times it may appear that particular species cannot invade a habitat because of the great efficiency or competitive ability of a species already in occupation; this is called "*competitive exclusion*".

A clear instance of competitive displacement is found in some of the prairie grasses of North America.[1] The most important native grass species of the north-west region is the bluebunch wheatgrass *Agropyron spicatum*. Towards the end of the 19th century a European grass species, cheatgrass or downy chess, *Bromus tectorum*, was introduced to the region, and quickly became dominant there, almost eliminating wheatgrass (and other species) from some areas. One reason why cheatgrass is so successful in competition with the native species is that it produces 65 to 200 times as many seeds as wheatgrass. Also, cheatgrass is an annual; it germinates and grows quickly. Wheatgrass, on the other hand, is perennial; each plant grows slowly over a number of years. But probably the most important reason for the success of cheatgrass is that its roots grow very rapidly, especially in the late winter. Experiments show that at this time of year its roots grow 50 per cent faster than those of wheatgrass. This is important since it enables cheatgrass to capture a bigger proportion of moisture during the dry season.

There are many other examples of the displacement of native species by an invader. The European starling *Sturnus vulgaris*, for instance,

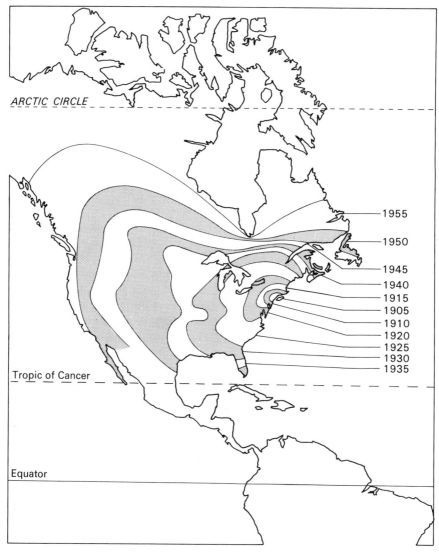

ARCTIC CIRCLE

1955
1950
1945
1940
1915
1905
1910
1920
1925
1930
1935

Tropic of Cancer

Equator

Figure 28. Gradual extension of the distribution of the European starling in North America from 1905 to 1955.

was introduced into Central Park, New York, in 1891. Since then it has spread widely and is now found in all of the States (Fig. 28). Mostly, it is found in urban areas and in the East has largely displaced the blue-bird *Sialia sialis* and the yellow-shafted flicker, *Colaptes auratus*. These species nest in tree-holes or in man-made holes, and starlings can occupy and hold most of the limited supply of these nest-sites. In the towns, then, the starling successfully competes with the native species for living space. But when flocks of starlings invade the countryside, they compete for food, insects, and seeds with the meadow larks (*Sturnella*) and these birds have also declined in some areas. On the other hand, some North American species have been successful in new habitats in Europe and Australia. An example is the American grey squirrel, *Sciurus carolinensis*, which was introduced into the British Isles in the 19th century. Between 1920 and 1925, the native red squirrel, *Sciurus vulgaris*, suffered a dramatic decline in numbers in Britain (largely due to disease following great abundance).[2] The spread of the grey squirrel has been accompanied by the disappearance of the red squirrel from many areas, particularly those in which their numbers were reduced by disease and those in which the grey squirrel first spread and established itself. This can be seen from the map (Fig. 29). In coniferous woodland, however, the red squirrel is apparently still the more successful, and the grey squirrel has rarely succeeded in occupying these areas. Where the grey squirrel has replaced the native red, it probably has done so by virtue of its superior adaptability to the niche of herbivore at canopy level in deciduous woodland.

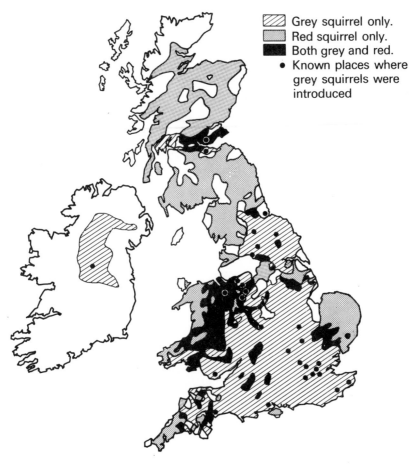

Figure 29. Map of squirrel distribution in Britain. (After Britain, Sunday Times Book Publications.)

There is good evidence that, in the past, whole faunas have invaded new areas and eliminated the native species by successful competition. For example, North and South America were separated by sea until the end of the Pliocene period (about two and a half million years ago). The Isthmus of Panama then came into being and enabled South American species to invade northwards and North American species southwards. In general, the northern species proved to be the more successful in competition and most of the characteristic South American fauna of this time became extinct; but very few North American species were wiped out by the South American invaders—probably because the North American herbivores were more efficient and the predators more successful.[3]

Despite these dramatic examples of invasion and competitive displacement it is most likely that in natural situations species that compete for food or other resources have evolved means of reducing the pressures of competition and of dividing up the resources between them. This is mutually advantageous since it reduces the risk of either species being eliminated and made extinct by competition with the others. This is an advantage not only to the species directly involved but to the whole community of species in the habitat since it results in *more* species, depending on as many different sources of food as possible. In such communities competition occurs between so many different species, each with its own specialized adaptations, that no single one can become so numerous as to displace others. This results in a greater degree of stability for the community, and stable communities are strongly resistant to the invasion of new species which might disrupt the highly-evolved pattern of competition within them. Because the community of species present in the early stages of ecological succession is a comparatively unstable one, new species can invade, so that changes in the species composition of the community take place.

Reducing competition Many different ways of reducing competition between species have evolved. Sometimes species with similar food or space requirements exploit the same resources at different seasons of the year, or even at different times of day. A common system amongst predatory mammals and birds is for one species (or a group of them) to have evolved specialized night-time activity whilst another species or group of species are day-time predators in the same habitats. An example amongst birds is the owls on the one hand, many species of which hunt at night, judging distance mostly by ear, and the hawks and falcons, on the other, which are daytime hunters, with extremely keen eyesight, especially adapted for judging distances accurately. Thus both groups of predators can co-exist in the same stretch of country, and prey on the same limited range of small mammals. Cases of this sort are described as *temporal separation* of species, and this is an effective method of eking out limited food resources amongst several species.

Probably much more common, however, are cases where the resources of a habitat are divided up between species by the restriction of each of them to only part of it, to specialized microhabitats. This is called *spatial separation* of species; it means that each species must be adapted to live in the fixed set of physical conditions of its particular microhabitat. It also means that such a species is not adapted to live in other microhabitats, and may find it difficult to invade them even if they were for some reason vacant and their food resources untapped.

An interesting example of spatial separation is found among barnacles on the rocky sea-shores of western Europe and north-eastern North America. Adult barnacles are firmly attached to rocks and feed when the tide is in by filtering plankton from the water. Two species that are common are *Chthamalus stellatus* which is found within an upper zone of the shore just below high-tide mark, and *Balanus balanoides* which occupies a much wider zone below that of *C. stellatus*, down to low-water mark (Fig. 30). The distribution of the two species does not overlap by more than a few centimetres. This situation was

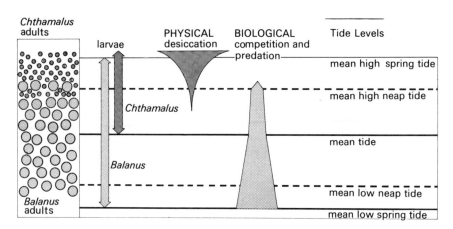

Figure 30. The distribution of the settling larvae and of the adults of the barnacles *Chthamalus* and *Balanus*, and the factors controlling this distribution. (After Odum, 1963.)

analysed by the ecologist J. H. Connell[4] who found that when the larva of *C. stellatus* ended their free-swimming existence in the sea and settled down for life they did so over the whole upper part of the rocks above mean tide level. The larvae of *B. balanoides* settled over the whole zone between high and low water including the area occupied by the adults of *C. stellatus*. Despite overlapping patterns of distribution of the larvae, the different distributions of the adults of the two species result from two separate processes. One acts on the *C. stellatus* zone at the top of the shore. The young *B. balanoides* are eliminated from this region because they cannot survive the long period of desiccation and the extremes of temperature to which they are exposed at low tide.

C. stellatus are more resistant to desiccation and survive. Lower down the rocks the *B. balanoides* persist because they are not exposed for so long, and here the larvae of *C. stellatus* are eliminated by direct competition from the young *B. balanoides*. These grow much faster and simply smother the *Chthamalus* larvae or even prise them off the rocks. Connell also performed experiments on these species and found that, if adult *B. balanoides* were removed from a strip of rock and young ones prevented from settling, the *C. stellatus* were able to colonize the full length of the strip right down to low tide level. This showed that the competition with *B. balanoides* was the main factor limiting the distribution of *C. stellatus* to the upper part of the shore.

This is an example of *competitive exclusion*, which in this case acts at the small-scale level within the species' habitat. But exclusion may also occur on a regional scale, limiting species to particular habitats, or on a geographical scale, limiting species to regions or continents. A good example of this sort is seen in two small herbaceous plants in the British Isles. One species, *Galium saxatile*, the heath bedstraw, is quite common in grasslands and heathlands on rather acid soils, especially sandy ones. The other, *Galium pumilum*, the slender bedstraw, is very closely related. Though both bear small white flowers during July and August, *G. pumilum* grows on alkaline soils on chalk or limestone, and it is sometimes easier to identify the species from the place where it is growing than by the appearance of its leaves and flowers. The British ecologist Sir Arthur Tansley performed experiments on the two species.[5] He found that in a greenhouse both species would grow in either acid or alkaline soils, although each species grew best on the type of soil in which it was most commonly found. He then grew a mixture of the two species on the two types of soil. He found the *G. pumilum* grew so much faster on alkaline soils that it crowded out *G. saxatile*, by filling the space available for roots and shading it from sunlight. Similarly *G. saxatile* was the more successful competitor on acid soils. Tansley suggested that the distribution of the two species could be largely explained by these competition effects, and by differences in their relative growth rates in the two types of soil. *G. saxatile* is absent even on some acid soils, perhaps because conditions are such that other vigorously-growing acid soil species have there been able to take its place.

Changing chaffinches Sometimes closely related species that have evolved in different geographical regions extend their geographical ranges till their distributions overlap. When such species have similar requirements for food and space, the pressure of competition may cause one or both species to change its habits. One such species is the chaffinch, *Fringilla coelebs*, which is found all over Europe and northeast Africa. It has very wide ecological tolerance and is one of the two or three most common birds over much of its range. It nests in a wide variety of habitats, including both coniferous and deciduous forest, as well as in parks and gardens. At some time in its history the chaffinch has invaded the Canary Islands, which lie one to two hundred miles off the north-west African coast. On two of the bigger islands, Grand Canary and Tenerife, the colonizing chaffinches have evolved into a new species, the blue chaffinch, *Fringilla teydea*. At some more recent time a second wave of colonists belonging to the parent species, *F. coelebs*, has reached the islands and, although they have not evolved into a new species, they have become a sub-species, *F. coelebs canariensis*. This form is forced to compete for food and nesting sites with the blue chaffinch.[6] In contrast to its ubiquitous distribution elsewhere, *F. coelebs* is found in these islands to breed only in the deciduous forests (which are mostly of sweet chestnuts and laurel), while the blue chaffinch occupies the coniferous forest. Clearly, the presence of the blue chaffinch restricts *F. coelebs* to those habitats in which it is the more effective competitor. This is confirmed by the fact that in the most westerly island, Palma, from which the blue chaffinch is for some reason absent, *F. coelebs* occupies the same range of forest habitats as in Europe.

Predators and prey Predators may be another biological factor influencing the distribution of species, but their effects have been much less studied than those of competition. The simplest influences that predators might have is to eliminate species by eating them, or alternatively, to prevent the entry of new ones into a habitat. There is very little evidence that either of these processes are common in nature. One or two experimental studies have shown that predators sometimes eat all the representatives of a species in their environment, particularly when the species is already rare. But all such studies have been made in rather artificial situations in which a predator is introduced into a community of species that have reached some sort of balance with their environment in the absence of the predator; such communities are not at all like natural communities which already include predators. In general, it is not in the interests of predatory species to eliminate a prey species because if they do this they destroy a potential source of food. Probably most natural communities have evolved so that there is a great number of potential prey species available to each predatory species. Thus no species is preyed upon too heavily, and the predators can always turn to alternative food species if the numbers of their usual prey should be reduced by climatic or other influences.

As mentioned earlier, competition may prevent two species from living together in a habitat, and may modify the distribution of species, because the resources of the habitat are inadequate to support both of them. Probably the most important effect of predators and of parasites and disease (which are "internal" predators) on the distribution of species is that, by feeding on the individuals of more than one species, they reduce the pressures of competition between them. Thus by reducing pressures on the resources of the habitat, predators may allow more species to survive than would survive if the predators were not there. Nearly 20 years ago it was shown by laboratory experiments that, if two species of seed-eating beetle (weevils) were kept together in jars of wheat, one species always eliminated the others within five generations. One of the two species always multiplied faster than the other, and this species won in the competition for food and places to lay eggs. But if a predator was introduced, such as a parasitic wasp, whose larvae feed inside the bodies of the beetle larvae of either species and eventually kill them, both species persisted. The numbers of both species were kept so low by the predator that competition for food, which would have caused one or other to be eliminated, never occurred.

More recent studies of natural communities have largely confirmed the hypothesis that predators may actually *increase* the number of different species that can live in a habitat. One especially fine study was made by the American ecologist Robert T. Paine on the animal community of a rocky shore on the Pacific coast of North America at Mukkaw Bay, Washington.[7] The community included 15 species, comprising acorn barnacles, limpets, chitons, mussels, dog whelks and one major predator, the starfish, *Pisaster ochraceus*, which fed on all the other species. Paine carried out an experiment on a small area of the shore in which he removed all the starfish and prevented any others from entering. Within a few months 60–80 per cent of the available space in the experimental area was occupied by newly-settled barnacles, which began to grow over other species and to eliminate them. After a year or so, however, the barnacles themselves began to be crowded out by large numbers of small, but rapidly-growing mussels, and when the study ended these completely dominated the community, which now consisted of only eight species. The removal of predators thus resulted in a decrease of nearly half in the number of species and there was evidence, too, that the number of plant species of the community (rock-encrusting algae) was also reduced, because of the competition with barnacles and mussels for the available space.

A general conclusion then is that the presence of predators in a well-balanced community is likely to increase rather than reduce the numbers of species present, so that, overall, predators broaden the distribution of species. Only a few experiments similar to Paine's have been performed and so one must be cautious about applying this conclusion to all communities. There is some independent evidence, however, that herbivores, which act on plants as predators do on their prey, may similarly increase the number of plant species that can live in a habitat. In the last century, Charles Darwin noticed that in southern England, meadowland grazed by sheep often contained as many as 20 species of plants, while neglected, ungrazed land contained only about 11 species. He suggested that fast-growing, tall grasses were controlled by sheep grazing in the meadow, but that in ungrazed land these species grew tall so that they shaded the small slow-growing plants from the sun and eliminated them.

On the Washington coast Paine performed another series of experiments in which he removed the sea-urchin *Strongylocentrotus purpuratus*, which grazes on algae.[8] Initially, there was an increase in the number of species of algae present: the six or so new species were probably ones that were normally grazed too heavily by the sea-urchin to survive in the habitat. But over two or three years the picture changed as the community of algae gradually became dominated by two species, *Hedophyllum sessile* on exposed parts of the shore, and *Laminaria groenlandica* in the more sheltered regions below low water mark. These two species were tall and probably "shaded-out" the smaller species, as did the tall grasses studied by Darwin. The total number of species present was in the end greatly reduced after the removal of the herbivores.

The activities of carnivorous predators in a community also have an effect on the plants since, by limiting to some extent the number of their herbivorous prey, they prevent over-grazing, and thus reduce the risk of rare species of plants being eliminated.

The niche   It seems, then, that competition for food or space, sometimes influenced by predators or grazing, is the main biological factor influencing the distribution of animal and plant species. The range of physical factors of the environment to which a species is adapted largely determines its range on a geographical or habitat scale, but within this range the survival of the species depends on its ability to compete for resources with other species living there. To understand the full significance of this for biogeography, some basic points of ecological theory must be considered. The first of these is the concept of the *ecological niche*. The ecological niche of a species (sometimes referred to simply as the "niche") is the way of life or "profession" that it practises in the particular part of the environment in which it lives. For instance, the niche of *Pisaster ochracreus*, the starfish studied by Paine, is that of the main predator of the rocky shores of the Pacific coast of North America. Similarly, the niche of the grey squirrel *Sciurus carolinensis*, is that of nut, seed, and catkin feeder of the canopy level of mainly deciduous woodland of the eastern United States (and now also of the southern British Isles). The niche of a species is defined then in terms of both its food and its habitat or microhabitat. It can be stated as a general principle that no two species can have the same ecological niche. When species compete for a niche, one will almost always be eliminated. This is the *Principle of Competitive Exclusion*, referred to earlier in the chapter, and originally suggested by the Russian biologist G. F. Gause[9] in the 1930's. Though it has since been realized that there are one or two "modifying clauses" which must be added to this principle, later work has shown it to be largely valid. Many cases have been observed where two or more species in a community seem at first sight to be occupying the same niche, but more detailed work has shown that they differ somewhat in their food habits, or in the area of the habitat in which they live. In some cases species may share part of their range of food or space with others, but usually each species has part of its niche, some food materials or a specialized microhabitat which is exclusively its own, and this acts as a refuge from competition with other species. No case of two species having the same niche has been demonstrated. A possible exception might arise in cases where the number of competing species is kept so low by a common predator that they never reach levels at which they might compete. This situation may be common, but no example has been studied in detail.

There have been some fine studies of the ways in which groups of closely related species with similar food or space requirements have evolved in such a way that a division of their resources into a series of distinct ecological niches has resulted, so that they do not compete with each other. One of the best of these was carried out by Robert H. MacArthur of Princeton University, who has contributed much to our understanding of the ecological niche.[10] In north-eastern North America five species of warbler often live together in coniferous forest. All belong to the same genus, *Dendroica*, and are very closely related;

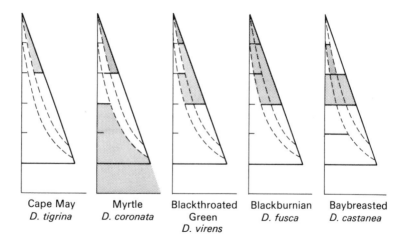

Cape May	Myrtle	Blackthroated	Blackburnian	Baybreasted
D. tigrina	*D. coronata*	Green	*D. fusca*	*D. castanea*
		D. virens		

Figure 31. Feeding zones on spruce trees of five species of North American warbler, *Dendroica*. The tree is shown diagrammatically, having been divided into six zones each 10 feet high, also into an outer zone of new needles and buds, a middle zone of old needles, and an inner zone of bare or lichen-covered branches. The shaded areas indicate regions where the species concerned spends at least half of its feeding time.

they all feed on insects and have similar requirements for nest sites. MacArthur found that the species showed complex differences in feeding and nesting behaviour which gave each one a distinct niche, and prevented competition, at least at times when food was plentiful. This is shown in Figure 31.

There are certain difficulties in applying the concept of the ecological niche to plants. Nearly all plants need sunlight for photosynthesis and, at least in terrestrial environments, they tend to occupy the same part of the habitat—the region just above and below the soil surface. Indeed to a great extent the plants actually form the microhabitats amongst which the animals of the community are distributed. Nevertheless, plant species do show differences in the areas they occupy horizontally and vertically in the habitat, and also in their seasons of growth and flowering. These differences are comparable to the ecological niches of animals, and there is little doubt that the Principle of Competitive Exclusion applies equally well to plants as to animals.

What are the factors that determine which species will occupy a niche when two or more species are competing for it? Experiments have shown that the species that wins the competition in a particular environment is the one best adapted to the range of physical conditions present. The winner is the species that can maintain the faster rate of population growth—the highest birth rate and the lowest death rate—under the particular conditions of the environment. In other words it is the species that can use the food resources and living-space most efficiently.

This is a most important conclusion for the biogeographer since it makes clear the difficulties that species have in broadening or changing their distributions. Simplifying somewhat, there are four possible results if a species is entering a geographical region or habitat which it has not previously inhabited.

First, the individuals of the species that are invading may be eliminated because they are not sufficiently well-adapted to the physical conditions of the region or habitat to compete successfully with species occupying suitable ecological niches. This is probably the case with the 50 or 60 species of birds that are "accidental" visitors to the British Isles. Many will leave as they have come, others will die of starvation and some may fall prey to predators. The end result is the same: that the species fails to establish itself permanently.

Second, the species may prove to be such a successful competitor that it is the native species, at present occupying a suitable niche, that is eliminated. This is what has occurred with the North American grasses mentioned earlier. The invader, *Bromus tectorum*, has eliminated *Agropyron spicatum* and other native species from parts of the United States.

Third, the invading species may prove to be an efficient competitor in parts of the niche of a native species and may succeed in displacing it from this part of the niche, which then becomes two smaller niches. Probably this is what has occurred in the case of the two species of chaffinch (discussed on page 65) in the Canary Islands. *Fringilla coelebs* has succeeded in displacing the blue chaffinch, *F. teydea*, from part of its niche in the islands where both species occur, and the original niche now consists of two niches, a coniferous one, occupied by *F. teydea*, and a deciduous forest one occupied by *F. coelebs*. A similar situation exists with the two species of squirrel in the British Isles.

Fourth, there are some circumstances in which an invading species is fortunate enough to find its normal niche vacant, either during the ecological succession or in the climax community, as follows.

In the course of ecological succession in a habitat, new ecological niches are constantly created by the activities of the species of the community themselves. The plants make available in their growth many new food sources and microhabitats for herbivores and the establishment of these in turn creates new niches for predators. In time the development of the community causes changes in the soil which for various reasons favour the growth of new species of plants which, when they invade, successfully compete with, and eliminate, the old ones. These again offer opportunities for the herbivores, so that the species of animals occupying the various niches change too. Eventually the *climax community* evolves and the creation of new ecological niches ceases. If a new species is to invade a climax community it must displace an existing species from its niche without the help of changes in the environment unfavourable to the current occupant. This is why climax communities are much more stable and less open to invasion than the earlier stages of succession.

Even in climax communities it is sometimes possible to recognize ecological niches that are vacant and would be available to an invader.[11] For instance, the tropical rain forest of the East Indies contains about 27 species of woodpeckers, which feed mostly on the insects in living or dead tree trunks and branches. The forest of New Guinea, only a few hundred miles to the east, contains not a single woodpecker. No native species have evolved to fill this niche, and apparently none of the East Indies species have invaded, so that the niche remains without an occupant. About 12 species of fish from the Red Sea have become established in the eastern Mediterranean after invading through the Suez Canal. They seem to have found unoccupied niches in the Mediterranean because several have become quite abundant (some have become important to the fishing industry) and yet no native species appears to have suffered.

It remains true, however, that there seems to be a limited number of niches available in each habitat or biome, even if a few do remain vacant, and competition for niches is the main biological factor limiting the distribution of species. One piece of evidence confirming that habitats contain a limited number of niches is that similar communities in different parts of a biome generally contain very similar numbers of species, with much the same range of foods and microhabitats. For example, a study of the types of feeding behaviour of birds in the rain-forest of Malaya and Australia[12] showed that there were fewer birds (about 117 species) in the Australian forest than in the Malayan ones (306 species). But, when the proportions of the species that are herbivorous, carnivorous, insectivorous (insect-feeding) and omnivorous (feeding on both plants and animal material) are compared (see Fig. 32), it is clear that the proportion of the species with particular feeding behaviour is very similar in the two areas. The correspondence is even more striking if the minor habitats, or vertical zones, of the forest are considered. For instance, in the Malayan forests, 40 per cent of the species of birds are insectivores occupying the middle zone of the

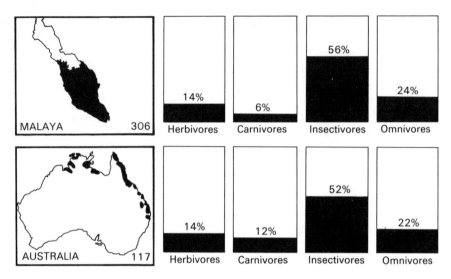

Figure 32. The feeding habits of birds in the rain forests of Malaya (top) and of Australia (bottom).

foliage, above the bare tree trunks, but below the upper parts of the canopy; in the Australian forest 42 per cent of the species are insectivores occupying this zone. Allowing for the possibility of a few unoccupied niches, it is likely that the number of species occupying a habitat or part of it is the same as the number of niches that it affords. Probably the Malayan and Australian forest present a similar range of niches for birds (as shown by the similar proportions using particular foods in particular parts of the habitat), but the total number of niches in the forest is greater in Malaya (as shown by the greater total number of species present).

Tropical environments generally contain many more species than temperate ones, implying that they contain more ecological niches. The reason for this is still unclear, though it has been suggested that two important factors may be the greater complexity of the tropical habitat, and its greater stability which has allowed gradual specialization to more restricted ecological niches. This greater species diversity of the tropical habitats is shown both by the total number of species of all kinds present and by the number of species of different groups of organisms present. Figure 33 shows how the number of species of birds breeding in various regions of North and Central America increases from the north to tropical regions.[13] Panama, only 500 miles north of the Equator, has about three times as many species (in a much smaller area) as Alaska. Similar differences between northern and southern regions are found in other groups of organisms; for instance, Labrador has 390 species of flowering plants and 169 species of beetles, while Florida has 2500 plants and 4000 beetles.

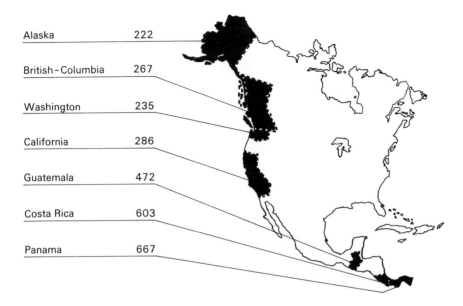

Alaska	222
British-Columbia	267
Washington	235
California	286
Guatemala	472
Costa Rica	603
Panama	667

Figure 33. Number of bird species in different parts of North and Central America.

Equivalent species   The same niche may often be occupied by different species in different areas of a single biome. In the grassland biome there is a niche for a fairly large herbivore, usually living in herds, and often adapted for fast running. In North America the niche is (or rather was) occupied by the bison in some regions and the prong-horn antelope in others, in Africa by many species of antelope, gazelles, the zebra and eland, in Eurasia by wild horses and asses, in South America by the pampas deer (*Odocoileus bezoarticus*) and the guanaco (*Lama huanacus*) and in Australia by kangaroos and wallabies. Such groups of species, occupying the same niche in different biomes, are called *ecologically equivalent species*. As these examples show, equivalent species are not necessarily closely-related, and quite different types of animals and plants may occupy the same niche in different places (such as the kangaroos in Australia and the ecologically equivalent pampas deer of South America). This is because species tend to evolve to fill particu-lar niches as a result of the pressures of competition, and sometimes of predation, in the community in which they live. Which species evolves to fill a particular niche is largely the result of historical and geographical accident. The development of adaptations enabling a species to occupy a particular niche enables it to escape from the pressure of competition that affects it elsewhere and results in the formation of a new species. This process of the evolution of species to fill niches that are available is called *adaptive radiation*. The way this occurs and the role of evolution in the distribution of species is the subject of the next chapter.

73

References

1 HANSON H.C. & CHURCHILL E.D. (1961) *The Plant Community.* Chapman & Hall, London; Reinhold, New York.

2 SHORTEN M. (1954) *Squirrels.* Collins, London.

3 SIMPSON G.G. (1969) South American mammals. In: Fittkau *et al.* (eds.), *Biogeography and Ecology in South America*, pp. 879–909. Junk, The Hague.

4 CONNELL J. (1961) The influence of interspecific competition and other factors on the distribution of the barnacle *Chthamalus stellatus. Ecology* **42**, 710–723.

5 TANSLEY A.G. (1917) On competition between *Galium saxatile* and *Galium sylvestre* on different types of soil. *J. Ecol.* **5**, 173–179.

6 LACK D. & SOUTHERN H.N. (1949) Birds on Tenerife. *Ibis* **91**, 607–626.

7 PAINE R.T. (1966) Food web complexity and species diversity. *Am. Nat.* **100**, 65–75.

8 PAINE R.T. & VADAS R.L. (1969) The effect of grazing in the sea urchin *Strongylocentrotus* on benthic algal populations. *Limnol. Oceanogr.* **14**, 710–719.

9 GAUSE G.F. (1934) *The Struggle for Existence.* Williams & Wilkins, Baltimore.

10 MACARTHUR R.H. (1958) Population ecology of some warblers of northeastern coniferous forests. *Ecology* **39**, 599–619.

11 MAYR E. (1966) *Animal Species and Evolution.* Harvard University Press, Cambridge, Mass.

12 HARRISON J.L. (1962) The distribution of feeding habits among animals in a tropical rain forest. *J. Anim. Ecol.* **31**, 53–63.

13 MACARTHUR R.H. & CONNELL J.H. (1966) *The Biology of Populations.* Wiley & Sons, New York and London.

CHAPTER 4

THE SOURCE OF NOVELTY

Some insects are protected from detection by predators by having an almost perfect resemblance to a leaf or a twig. This is perhaps the most dramatic example of the intricate way in which an organism is adapted to its environment. Other adaptations are just as intricate and thorough, although not so obvious. Every aspect of the environment makes its demand upon the structure or the physiology of the organism: the average state of the physical conditions, together with their daily and annual ranges of variation; the changing patterns of supply and abundance of food; the occasional increased losses due to disease, to predators, or to the increased competition from other organisms at the same level in the ecological food web. Every species of animal or plant must be adapted to all these conditions; it must be able to tolerate and survive the hostile aspects of its environment, and yet able to take advantage of its opportunities.

Before Darwin put forward his theory of evolution by natural selection, it was accepted that each species that we see today had always existed precisely as we now see it. God had created each one, with all its detailed adaptations, and these had remained unchanged. Fossils were merely the remains of other types of animal, each equally unchanging during its span of existence, which God had destroyed in a catastrophe (or a number of catastrophes) such as the biblical Flood.

In his journey round the world in the ship H.M.S. *Beagle* from 1831 to 1836, Darwin saw two phenomena that eventually led him to disbelieve all this. On the Galapagos Islands in the Pacific, isolated from South America by 600 miles of sea (Fig. 34), different birds were well adapted to feeding on different diets. Some, with heavy beaks, cracked open nuts or seeds; some, with smaller beaks, fed on fruit and flowers; others again, with fine, narrow beaks, fed on insects. On the mainland, these different niches are occupied by quite different, unrelated types of bird—for example, by toucans, parrots and flycatchers. The remarkable fact was that on the Galapagos Islands each of these varied niches was instead filled by a differently adapted species of one type of bird, the finch. It looked very much as though finches had managed to colonize the Galapagos Islands before other types of bird and then, free from their competition, had been able to adapt to diets and ways of life that were normally not available to them. This logical explanation, however, ran directly against the current idea of the fixity of characteristics. Equally disturbing were the fossils which Darwin had found in South America. The sloth, armadillo and guanaco (the wild ancestor of the domesticated llama) were each represented by fossils which were larger than the

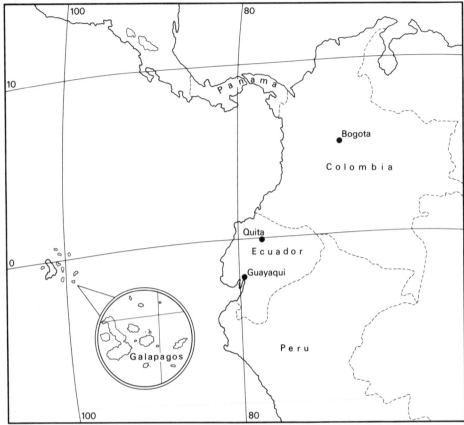

Figure 34. Map of Pacific showing position of Galapagos Islands in relation to South America.

living forms, but were clearly very similar to them. Again, the idea that the living species were descended from the fossil species was a straight-forward explanation, but one that contradicted the view that each species was a special creation and had no blood relationship with any other species.

Natural selection The explanation that Darwin eventually deduced and published in 1858 is now an almost universally accepted part of the basic philosophy of biological science. Darwin realized that any pair of animals or plants produces far more offspring than would be needed simply to replace that pair: there must, therefore, be competition for survival amongst the offspring. Furthermore, these offspring are not identical with one another, but vary slightly in their characteristics. Inevitably, some of these variations will prove to be better adapted to the mode of life of the organism than others. The offspring which have these favourable characteristics will then have a natural advantage in the competition of life, and will tend to survive at the expense of their less fortunate relatives. By their survival, and eventual mating, this process of *natural selection* will lead to the persistence of these favourable characteristics into the next generation.

Evolution is therefore possible because of competition between individuals that differ slightly from one another. But why should these differences exist, and why should each species not be able to evolve a single perfect answer to the demands that the environment makes upon it? All the flowers of a particular species of plant would then, for example, be of exactly the same colour and every sparrow would have a beak of precisely the same size and shape. Such a simple solution is not possible, because the demands of the environment are neither stable nor uniform. Conditions vary from place to place, from day to day, from season to season. No single type can be the best possible adaptation to all these varying conditions. Instead, one particular size of beak might be the best for the winter diet of a sparrow, while another might be better adapted to its summer food. Since, during the lifetimes of two sparrows differing in this way, each type of beak is slightly better adapted at one time and slightly worse adapted at another, natural selection will not favour one at the expense of the other. Both types will therefore continue to exist in the population as a whole.

Because we do not normally examine sparrows very closely, we are not aware of the many ways in which the individual birds may differ from one another. In reality, of course, they vary in as many ways as do different individual human beings. In our own species, we are accustomed to the multitude of trivial variations that make each individual recognizably unique—the precise shape and size of the nose, ears, eyes, chin, mouth, teeth, the colour of the eyes and hair and the type of complexion, the texture and waviness of the hair; the height and build, the pitch of voice, and degree of resistance to different stresses and diseases. We know of other, less obvious characteristics in which individuals also differ, such as their fingerprints and their blood group. All of these variations are, then, the material upon which natural selection can act. In each generation, those individuals that carry the greatest number of less advantageous characteristics would be least likely to live long enough to have children who would perpetuate these traits, while those with a large number of more advantageous characteristics would be more likely to survive and breed successfully.

Changes of this kind in the characteristics of a species are not merely theoretical deductions, but can be shown to have taken place. This can be seen clearly only when the environment of a species has changed rapidly, which does happen, though rarely. A particularly clear example resulted from the darkening of the countryside around the industrial cities of Great Britain during the second half of the nineteenth century.[1] This change greatly affected those moths, such as the peppered moth, *Biston betularia*, which relied upon camouflage to protect them from being seen and eaten by insectivorous birds. As long as the bark of the trees on which they rested was pale, it was advantageous for the moths to be pale also. But as industrialization proceeded and the bark of trees

Figure 35a. Photograph showing the inconspicuousness of the normal form and the conspicuous appearance of the melanic form of the peppered moth (*Biston betularia*) on a normal, lichen covered tree trunk.

near the cities became blackened by soot, the pale individuals of *Biston betularia* were now more and more conspicuous (Fig. 35). It was no coincidence that it was in 1848 that a dark or "melanic" form of this moth first appeared and gradually became more and more common in these industrial areas until, by 1895, it was the pale form which was now the rare exception near the cities. The reason for this change is quite clear. Against the soot-darkened bark it was now the melanic form which was less conspicuous and therefore favoured by natural selection. Experiments have confirmed this deduction. A large number of peppered moths, some light and some dark, each marked by a tiny spot of paint, were released in two areas. In the first, a non-industrial area, later trapping led to the recapture of 14·6 per cent of the pale coloured moths, but of only 4·7 per cent of the melanic forms—far more of these

Figure 35b. The reverse situation when the same forms are on soot-covered, lichen-free bark. Photographs by S. Beaufay, from Wallace and Srb, *Adaptation* (Prentice-Hall, Inc.).

had already been eaten by birds. In the industrial area the proportions were reversed: only 14 per cent of the pale moths were recaptured, but 27·5 per cent of the dark forms, which here had been better camouflaged than their paler relatives. It is also interesting to find that, near towns which have in recent years taken measures to reduce smoke production, the proportion of melanic moths has already dropped.

This is not the only known example of a rapid change in the characteristics of an insect, due to the influence of man. The use of DDT to control insects conferred a great advantage upon those which were resistant to this chemical, and a high proportion of house flies are now of the DDT-resistant variety. The evolution of strains of bacteria resistant to commonly used antibiotics provides other clear examples of Darwin's principle of natural selection in action.

If all the members of a species gradually came to possess such new characters as a resistance to DDT, the species would have changed but no additional species would have resulted. However, as Darwin realized in the Galapagos Islands, the original single species may also split into two or more new species. In order to explain how this happens, the meaning of the term "species" must first be explained. Why do biologists consider that a sparrow and a robin are separate species, but that an alsatian and a greyhound (which appear just as different from one another) are both members of the same species? To biologists, the essential difference is that, under normal conditions in the wild, a sparrow and a robin do not mate together, while an alsatian and a greyhound will (the great difference in the appearance of the two dogs is due to artificial selection by man). Sometimes the difference is a little more subtle, as in the case of the horse and the ass; though these are separate species, they do sometimes breed together. However, this cross between the two species is short-lived and does not result in a permanent merging, because the resulting mule or hinny is sterile.

There are, then, two phenomena that have to be explained: the splitting of one species into two, and the inability of these two species to breed together. As will be seen, these are really two aspects of a single problem, and a clue has already been provided by the melanic form of the peppered moth. This is the common variety around the industrial cities of Britain where experiments showed it was better adapted than the pale variety, while the reverse is true in country districts far from the cities. Within the species as a whole, more than one type of adaptation is now found, but the distribution of each is distinct. The process of adaptation has taken place independently in each population, not in a general way throughout the entire species.

This independent adaptation of each population is possible because each is, to some extent at least, isolated from other populations. Though the distribution of the peppered moth covers the whole of the mainland of Great Britain, in fact the moth is found almost exclusively within patches of light woodland. It is only found in the intervening areas of open country if high winds have blown it from its normal woodland habitat. As a result, the peppered moths of each patch of woodland will, over long periods of time, mate only with one another. This, then, is the source of the isolation that permits the independent adaptation of each population.

Each species is broken up in this way into separate populations, cut off by areas in which physical or biological factors make it difficult or impossible for the species to survive. As long as the isolation persists, each population will gradually tend to become slightly different from the others, due to the action of two forces. One of these forces lies within the animal or plant, the other lies outside it.

The isolating force within the organism The force within the organism lies in the system that is responsible for the transmission of the characters of the parents to the next generation. Within each cell lies a rather opaque object called the *nucleus,* inside which is a number of thread-like bodies called *chromosomes.* These chromosomes consist of a chain of large complex molecules known as *genes.* It is the bio-chemical action of these genes that is responsible for the characteristics of every cell of an individual, and thus for the characteristics of the organism as a whole. There might, then, be a particular gene that determined the colour of an individual's hair, while another might be responsible for the texture of the hair and another for its waviness. Each gene exists in a number of slightly different versions, or *alleles.* Taking the gene responsible for hair colour as an example, one allele might cause the hair to be brown while another might cause it to be red. Many different alleles of each gene may exist, and this is the main reason for much of the variation in structure that Darwin noted.

An individual of course inherits characteristics from both of its parents. This is because each cell carries not one set of these gene-bearing chromosomes, but two: one set derived from the individual's mother, and the other derived from its father. A double dose of each gene is therefore present, one inherited from the mother, the other from the father. Both parents may possess exactly the same allele of a particular gene. For example, both may have the allele for brown hair, in which case their offspring would also have brown hair. But, very often, they may hand down different alleles to their offspring; for example, one might provide a brown hair allele, while the other provided a red hair allele. In such a case, the result is *not* a mixing or blurring of the action of the two alleles to produce an intermediate such as reddish-brown hair. Instead, only one of the two alleles goes into action, and the other appears to remain inert. The active allele is known as the *dominant* allele and the inert one as the *recessive* allele. Which allele is dominant and which recessive is normally firmly fixed and unvarying—in the hypothetical example given, the brown hair allele might be dominant, and the red hair allele might be recessive.

The genes themselves are highly complex in their biochemical structure. Though normally each is precisely and accurately duplicated each time a cell divides, it is not surprising that from time to time—due to the incredible complexity of the molecules involved—there is a slight error in this process. This may happen in the cell divisions which lead to the production of the sexual gametes (the male sperm or pollen, the female ovum or egg). If so, the individual resulting from that sexual union may show a completely new character, unlike those of either of its parents. In the example given above, such an individual might have completely colourless hair. Such sudden alterations in the genes are known as *mutations.*

The genetic system outlined above can lead to changes in the characteristics of an isolated population in two ways. Firstly, new mutations may appear and spread through the population. Secondly, since each individual carries several thousand genes, and each may be present in any one of its several different alleles, no two individuals carry exactly the same genetic constitution, or *genotype*—unless they are identical twins, developed from the splitting of a single original developing egg. Inevitably, therefore, the two isolated populations will differ somewhat in their initial genetic content, some alleles being rarer in one population than in the other or, in extreme cases, being absent altogether. As mating goes on in the two populations, new combinations of alleles will appear haphazardly in each, and this will lead to further differences between them.

External isolating forces It is, then, the independent appearance of new mutations in each population, and the independent course of genetic change within populations, that together make up the driving force within the organism which tends to make each isolated population gradually become different from every other. The force *outside* the organism that aids the process is simpler. Natural selection acts to adapt the population to its surroundings. But no two patches of woodland, no two freshwater ponds, will be absolutely identical, even if they lie in the same area of country. They may differ in the precise nature of their soil or water, in their range of temperature, or their average temperature, or in the particular species of animal or plant that may become unusually rare or unusually common in that locality. Since each population has to adapt to slightly different conditions, the two populations will gradually come to differ from one another.

The history of a patch of sunflowers living in a ditch in the Sacramento Valley of California provides a good example of the way in which all these forces can gradually make two populations become quite different from one another.[2] The population consisted of natural hybrids between the two annual sunflowers of California, *Helianthus annuus* and *H. bolanderi*. To begin with, the original population gradually became split into two by a drying-out of part of the ditch, the dry section being colonized by grasses among which the sunflowers could not survive. Over the space of five years, the dry grassy patch widened until it had pushed the two separated sub-populations of sunflowers over 100 metres apart. One of these was now in a deeper part of the ditch, which remained wet until late spring, while the other grew in a shallower, drier position. The two sub-populations became different in a number of characteristics, such as the shape of the flower head as a whole, the number of sterile floret rays surrounding the head, the shape of the base of the leaf, and the length of the hairs on the stem and leaves. Even though bees could easily fly from one population to the other, so that some cross-pollination between them must have taken place, observations over the next seven years showed that the differences between the two populations did not disappear. Their environments differed sufficiently to ensure that natural selection preserved the distinctiveness of the two populations.

Exactly the same process takes place in animal populations though, because they can move, the separate populations may each cover a

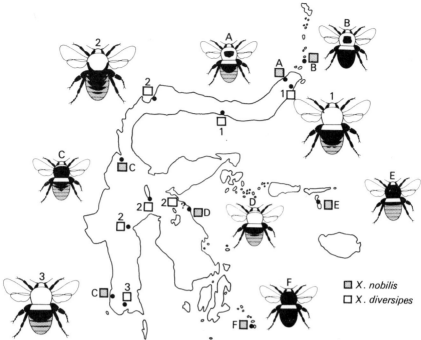

Figure 36. Colour patterns of two species of Carpenter Bee, *Xylocopa nobilis* (A–F) and *X. diversipes* (1–3) on the East Indian island of Celebes and neighbouring islets. Three separate populations of each species occur on the main island, each with distinct patterning. A further three pattern types of *X. nobilis* occur on the surrounding islets.

larger area. For example, two species of Carpenter Bee, *Xylocopa diversipes* and *X. nobilis*, live in the large East Indian island of Celebes.[3] Three different colour patterns of each species are known on Celebes, each making up one or more separate populations. Another three colour patterns of *X. nobilis* are known to exist on neighbouring islets (Fig. 36).

Once populations have started to diverge in their genetic adaptations in this way, the foundations for the appearance of a new species have been laid. If two divergent populations should meet again when the process has not gone very far, they may completely hybridize and merge into one another. The further, vital step towards the appearance of a new species is when hybrids between the two independent populations do appear, but only along a narrow zone where the two populations meet. Such a situation suggests that, though continued interbreeding within this zone can produce a population of hybrids, these hybrids cannot compete elsewhere with either of the pure parent populations. This seems to be the situation with the woodpecker-like flickers of North America. The eastern yellow-shafted flicker, *Colaptes auratus*, does mate with the western red-shafted flicker, *C. cafer*, along a narrow 2000-mile-long stretch where the two meet, but this zone of hybridization does not seem to be spreading (Fig. 37).

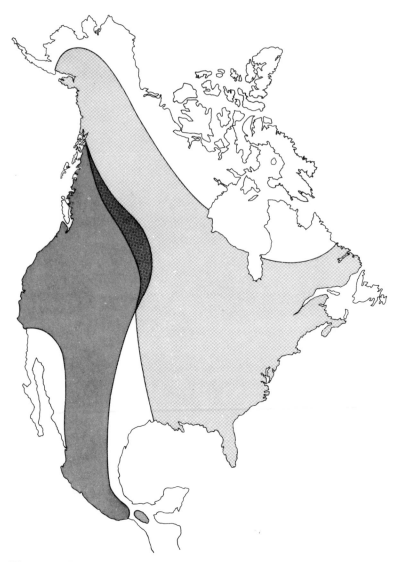

Figure 37. The Yellow-shafted flicker of eastern and north-western N. America hybridizes over a long but narrow zone with the Red-shafted flicker of the south-west.

Barriers to interbreeding   Once independent evolution in isolation has produced a situation in which the hybrids are less well-adapted than either of their parents, then natural selection will favour individuals that do not perpetuate this more poorly adapted hybrid population. This may be either because they cannot, or will not, mate with individuals from the other group, or because such unions are infertile. The barrier to hybridization is known as an *isolating mechanism*, and it may take many forms. In animals, such as birds and insects, that have complicated courtship and mating behaviour, small differences in these rituals may in themselves effectively prevent interbreeding. Sometimes the preference for the mating site may differ slightly. For example, the North American toads *Bufo fowleri* and *B. americanus* live in the same areas, but breed in different places.[4] *B. fowleri* breeds in large, still bodies of water such as ponds, large rainpools and quiet streams, whereas *B. americanus* prefers shallow puddles or brook pools. Interbreeding between the two species is also hindered by the fact that *B. americanus* breeds in the early spring and *B. fowleri* in the late spring —though there is some mid-spring overlap.

Many flowering plants are pollinated by animals that are attracted to the flowers by their nectar or pollen. Hybridization may then be prevented by the adaptation of the flowers to different pollinators. For example, differences in size, shape and colour of the flowers of related species of the North American beard-tongue (*Pentstemon*) adapt them to pollination by different insects—or, in one case, by a humming-bird (Fig. 38). In other plants, related species have come to differ in the time at which they shed their pollen, thus making hybridization impossible.

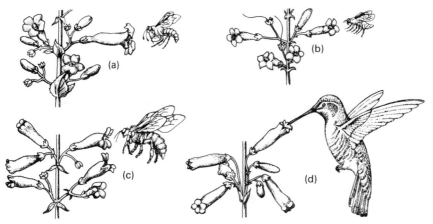

Figure 38. Four species of the Beard-tongue *Pentstemon* found in California, together with their pollinators. Species (a) and (b) are pollinated by solitary wasps, species (c) by carpenter bees, and species (d) by hummingbirds. (After Stebbins.)

Even if pollen of another species does reach the stigma of a flower, in many cases it is unable even to form a pollen-tube, presumably because the biochemical environment in which it finds itself is too alien. It cannot, therefore, grow down to fertilize the ovum. Similarly, in many animals alien spermatozoa cause an allergy reaction in the walls of the female genital passage and the spermatozoa subsequently die before fertilization.

Other isolating mechanisms may not prevent mating and fertilization taking place, but instead ensure that the union is sterile. These may be genetic isolating mechanisms, the structure and arrangement of the genes on the chromosomes being so different that the normal processes of chromosome splitting and pairing that accompany cell division are disrupted. These differences may make themselves felt at any stage from the time of fertilization of the ovum, through all the steps in development, to the time at which the sexual gametes of the hybrid itself are produced. Whenever the effects are felt, the result is the same: the hybrid mating is sterile or, if offspring are produced, these are themselves sterile (as in the case mentioned earlier of mating between a horse and an ass), or of reduced fertility.

Polyploids Another method by which new species can appear is by *polyploidy*—the doubling of the whole set of chromosomes in the nucleus of a developing egg or seed, so that each automatically has an identical partner. This may occur in the development of a hybrid individual (in which case it can overcome any genetic isolating mechanisms), or in the development of an otherwise normal offspring of parents from a single species. In either case, the new polyploid individual will be unlikely to find another similar individual with which to mate, and the origin of new species by polyploidy has therefore been important only in groups in which self-fertilization is common. Only a few animal groups fall into this category (e.g. turbellarians, lumbricid earthworms and weevils), but in these groups an appreciable proportion of the species probably arose in this way. In plants, however, in which self-fertilization is common, polyploidy is an important mechanism of speciation.[5] More than one-third of all plant species have probably arisen in this way, including many valuable crop plants such as wheat, oats, cotton, potatoes, bananas, coffee and sugar cane.

Summary Whatever may be the nature of the isolating mechanism that keeps them separate, two groups have become two independent species as soon as they are no longer able to interbreed or to produce fertile hybrids. Where only one species existed before, there are now two. They may have become sufficiently different in their adaptations to be able to spread into one another's area and co-exist without competing. For example, in North America two closely related species of bird, the Song Sparrow *Melospiza melodia* and the Lincoln Sparrow *M. lincolni* live together, as do the red maple tree *Acer rubrum* and the sugar maple tree *A. saccharum*.

The whole process of *speciation* (as the evolutionary process leading to new species is termed) is, then, able to start only because, since organisms can only exist under a limited range of conditions, and the conditions in the environment vary in an irregular fashion, each species becomes broken up into separate populations. Within each of these, new features are continually appearing due to genetic changes, and natural selection is constantly weeding out those new features least suited to the environment.[6]

Adaptations for survival The complex chemicals of an animal or plant cell are quite unlike most inorganic substances, and their complicated interactions can only take place within a limited range of physical conditions. In order to survive, therefore, any cell must continually ensure that its chemicals remain isolated from those of its surroundings, and that it remains within the range of conditions in which its own chemicals can continue to function. Evolutionary history has been the gradual process by which organisms have become able to isolate themselves (or, more precisely, their body tissues) from their surroundings with increasing effectiveness. This has made it possible for organisms to become able to survive in conditions that are more and more unfavourable for life—the conquest of dry land being a major step in that direction.

Alongside the evolution of adaptations providing insulation from the physical environment, organisms have also had to cope with difficulties due to their companion species. Both the existence of animals or plants that are similar to one another in their adaptations, and which therefore compete with one another, and the complicated interactions of herbivore and plant food or of predator and prey, lead to the appearance of new difficulties. Evolution is, therefore, the process by which organisms have conquered two types of barrier: those imposed by physical conditions, and those resulting from the biological world of animals and plants among which they live.

Even in the most favourable habitats, the physical conditions are rarely ideal for the organism through the 24-hour daily cycle. In addition to the daily alteration of light and dark, with the accompanying rise and fall in temperature and relative humidity, the temperature and rainfall may vary considerably from one day to another in many areas. Any organism must be able to tolerate changes of this kind in the physical conditions: animals may take shelter during rainstorms while plants may close their leaves or flowers.

As long as these conditions are short-lived, evasion or toleration is not difficult. A more serious problem exists in parts of the world where the climate is seasonal. Here the conditions may not be severe in themselves, but their continuation over a period of months demands a quite different adaptation on the part of the organism. Such prolonged alterations in physical conditions inevitably affect the whole community: the lower temperatures and shorter daylight hours of winter directly affect both plants and animals and, in addition, most animals find that food is then far less plentiful. Some animals, of course, are able to avoid these conditions altogether, by migrating to warmer climates. This solution is particularly common in flying animals, such as birds and butterflies, that can cover relatively long distances with ease, and to which a river or a stretch of sea is not an impassable obstacle. Other animals, such as bears and many smaller mammals, endure the cold and the scarcity of food by *hibernating*—reducing their metabolism to a minimum and surviving on food reserves they have stored up in the body during the summer. Resting stages, of one kind or another, are

87

common in both animals and plants of higher latitudes. The hard, resistant seed-cases of many plants, which will not germinate until they have been exposed to the coldness of winter (during which the parent plant may die), have their counterpart in the periods of arrested development of many insects. The hard, resistant chrysalis of a butterfly is a stage during which the complex changes from the caterpillar to the winged adult are carried out. But this inert, non-feeding stage is equally a convenient form in which, by a slowing-down of the rate of these changes, the whole winter can be passed in comparative safety.

For plants, the dry season of areas closer to the equator brings the risk of desiccation due to lack of water, which may also be unavailable during winter in cold-temperate latitudes because it is frozen into ice. Winter also brings the danger of frost damage. Though the stem of the plant can be protected by bark, the leaves are still exposed. The flowering plants were able to solve this problem by developing the mechanism of leaf-fall, so that the enormous, exposed leaf-surface is shed completely until the following spring. It is interesting to find that this adaptation was probably first developed by flowering plants in the tropical regions to reduce water loss during the dry season.[7] Only later did flowering plants with this *deciduous* habit spread to the colder regions where water is also in short supply because it is frozen. In addition to cutting down the rate of water loss, the reduction in the exposed surface area, due to the shedding of leaves in winter, reduces the damage caused by high winds and by settling snow in these regions.

Meeting the challenge of the environment The adaptations considered so far are all ones that in one way or another *evaded* the challenge of inhospitable conditions. To live and carry on all its normal functions in an area that is particularly cold, or particularly hot and dry, requires a more thoroughgoing adaptation of the whole organization of the organism and of its life-history. For example, most frogs and toads cannot survive in desert regions because the adults quickly become desiccated and because the water that the embryos need for their development rapidly becomes too hot for their existence and eventually evaporates completely. Nevertheless, some frogs and toads have been able to adapt to these conditions. An example is the spadefoot toad *Scaphiopus couchii* which is found in the deserts of the south-western United States.[8] Its eggs are laid in temporary desert rainpools resulting from local storms and their rate of development is very high. As a result, they pass through the most temperature-sensitive stages of their development before the afternoon of the next day, when the temperature rises above the level (about $34°c$) that is critical for them. The larvae also hatch at an early stage from the mass of jelly which surrounds them and are therefore soon able to seek the coolest parts of the rainpool. The adults can survive in these deserts because their hind limbs are modified to form scoop-like spades, with which they can excavate holes. In the hottest periods they can therefore retreat from the desert surface to the cooler, moister environment of their hole.

Even the spadefoot toad is, in retreating to its hole, still only temporarily able to evade the problems posed by the physical conditions of its environment; it will eventually have to emerge to feed. The limitations of this type of solution are as obvious as are the advantages of more fundamental adaptations that permanently insulate the organism from unfavourable physical conditions. The insulating coat, formed by hair in mammals, or by feathers in birds, helps them to maintain the internal temperature of their body at a constant level even if, like a camel or an ostrich, they live in a desert where the day-time temperature may be as high as $55°C$. The insulating coat also reduces their rate of water loss to a tolerable level, as does the resistant external skeleton of insects. Similarly, the thick cuticle of the leaves of evergreen plants and of conifer needles protects them from winter conditions which other flowering plants can only survive by shedding their leaves. Thus the evergreens are able to photosynthesize all year round, while deciduous species cannot manufacture food during the winter months when they have no leaves.

Competition for life Adaptations of this kind do not merely allow their owners to colonize regions having extreme climates. Since their internal conditions are more constant, the rates of their biochemical processes and the level of activity of these organisms can remain constant, irrespective of daily or seasonal climatic changes. Their comparative insulation from the effects of physical conditions therefore provides them with a considerable advantage in their competition with other organisms. It is no coincidence that the groups that have this insulation—the insects, birds and mammals—are by far the most numerous, varied and widespread of terrestrial animals.

However successful their adaptations to their physical environment, organisms must also adapt to the demands of the biological world around them, either to avoid being eaten, or to compete for space or food supply with other organisms. There can be no final solution to any of these problems for, as quickly as new adaptations appear that reduce predation or allow more successful competition with other species, the predator or competitor will in its turn adapt. The herbivorous group that becomes able to run faster and escape from its predators itself provides the stimulus that leads to the evolution of faster predators. The plant that evolves spines or unpleasantly tasting biochemicals to avoid being eaten by herbivores similarly stimulates the appearance of herbivores insensitive to these defences.

One method by which the problem of competition can be at least reduced is for the two competing groups gradually to become specialized to different ways of life; they may then be able to exist together in the same area without competing with one another. This is exemplified by the chaffinches of the Canary Isles discussed in Chapter 3.

Finally, a part of the adaptation of any population is to ensure that its numbers are approximately adjusted to the food supply of the area. The territorial behaviour of some birds, such as the Scottish red grouse, *Lagopus lagopus scoticus*, does this very effectively.[9] Each male takes possession of an area of heather moor large enough to provide an adequate food supply for its family, and defends it against other members of its species. In a year when food is scarce, the territory claimed is larger. The males compete for these territories by display, and this system therefore not only ensures that it is the weakest birds that are excluded from the moor (and are frequently killed by predators or starvation), but also ensures that an adequate food supply is available for the successful birds. This type of social competition is a close parallel to that in human societies. In both, as a result of social competition those which are successful receive a variety of advantages—sexual, nutritive and environmental. The red grouse society no longer contains a group of moderately successful males, sometimes adequately fed and at other times weakened by malnutrition. Instead, it is permanently divided into the "haves", assured of the necessities of life and of the opportunity, by reproduction, to transmit their characteristics to the next generation, and the "have-nots", of whom about 60 per cent die during the winter.

In many animals in which pairs do not occupy separate territories, there are daily visual or vocal displays of the whole population, especially at dawn or dusk. Certain ecologists have suggested that this is a method by which the population of a species becomes aware of its own density.[10] If this is unduly high, this might lead to physiological or behavioural changes, such as a reduction in the number of individuals that mate, or a reduction in the number of eggs laid, which could eventually reduce the size of the population. The absence of a system of this kind in our own species may well be one of the reasons why our numbers have so enormously increased that we now threaten to overwhelm the resources of our planet.

References

1 KETTLEWELL H.B.D. (1961) The phenomenon of industrial melanism in Lepidoptera. *Ann. Rev. Entomol.* **6,** 245–262.
2 STEBBINS G.L. & DALY K. (1961) Changes in the variation pattern of a hybrid population of *Helianthus* over an eight year period. *Evolution* **15,** 60–61.
3 VAN DER VECHT J. (1953) The carpenter bees (*Xylocopa* Latr.) of Celebes. *Idea* **9,** 57·59.
4 BLAIR A.P. (1942) Isolating mechanisms in a complex of four species of toads. *Biol. Symp.* **6,** 235–249.
5 STEBBINS G.L. (1950) *Variation and Evolution in Plants*. Columbia University Press, New York.
6 MAYR E. (1970) *Populations, Species and Evolution*. Oxford University Press, London.
7 AXELROD D.I. (1966) Origin of deciduous and evergreen habits in temperate forests. *Evolution* **20,** 1–15.

8 ZWEIFEL R.G. (1968) Reproductive biology of anurans of the arid South-West, with emphasis on adaptation of embryos to temperature. *Bull. Am. Mus. nat. Hist.* **140,** 1–64.

9 WARSON A. (1966) Social status and population regulation in the red grouse (*Lagopus lagopus scoticus*). *Proc. R. Soc. Pop. Study Group* **2,** 22–30.

10 WYNNE-EDWARDS V.C. (1962) *Animal Dispersion in Relation to Social Behaviour*. Oliver & Boyd, Edinburgh and London.

CHAPTER 5

LIFE ON ISLANDS

It is not surprising that it was in the minds of Darwin and Wallace, both students of island faunas, that the modern theory of evolution by natural selection first appeared. As has been seen in Chapter 4, evolution is likely to take place most rapidly in a small, isolated population—and such a population is more likely to be found on an island than anywhere else. Because of their isolation, islands also provide many fascinating examples of the ways in which some organisms, because of their structure, their physiology or their ecological preferences, have been unable to colonize a particular island, and of the ways in which other organisms have, by evolutionary change, occupied their niche.

Problems of access Oceans are the most effective barrier to the distribution of all land animals except those which can fly. Some flying animals, such as larger birds and bats, may be capable of reaching even the most distant islands unaided, using their own powers of flight—especially if, like water birds, they are able to alight on the surface of the water to rest without becoming waterlogged. Smaller birds and bats and, especially, flying insects may reach islands by being carried passively on high winds. These animals may, in their turn, carry the eggs and resting stages of other animals, as well as the fruits, seeds, and spores of plants, attached to their bodies. Many fruits and seeds have special sticky secretions or hooks to make them adhere to the bodies of animals. Examples are the spiny fruits of burdocks and beggarticks, and the berries of mistletoe, which are filled with a sticky juice so that the seeds they contain stick to birds' beaks. The winds themselves may also carry seeds or spores for great distances. Some plants have seeds that are specially adapted to being carried by the wind. Orchid seeds, for example, are surrounded by light, empty cells, and some have been known to travel over 200 km. Rhododendron and maple seeds have wings, and the seeds of many members of the Compositae (daisies and their relatives) have tufts of fluffy hairs—those of thistles have been carried by the wind for 145 km. The spores of most ferns and lower plants are so small ($\frac{1}{10}-\frac{1}{100}$ mm) that they are readily carried indefinite distances by winds.

Most land animals cannot survive in sea-water for long enough to cross oceans and reach a distant island, but it seems possible that some may occasionally make the journey on masses of drifting débris. Natural rafts of this kind are washed down the rivers in tropical regions after heavy storms, and entire trees may also float for considerable

Figure 39. Coconut palms on the edge of a tropical beach.

distances. Small animals such as frogs, lizards, and rats, may be carried from island to island in this way, while the resistant eggs of many other animals may find safe lodgement in this vegetation. It may seem unlikely that an animal could be transported in this way and arrive safely. However, even if the odds that this happens in any one year are as low as one in a million, then over the 65 million years of the Cenozoic Era, the odds are 65 to one *in favour* of successful colonization.

A few plants have developed fruits and seeds that can be carried unharmed in the sea. For example, the coconut fruit can survive prolonged immersion and the coconut palm (*Cocos nucifera*) is widespread on the edges of tropical beaches (Fig. 39). But, since the beach is as far as most seaborne fruits or seeds can hope to get, it follows that only species that can live on the beach are able to colonize distant islands in this way. The fruits or seeds of plants that live inland would be less likely to reach the sea and, even if they were able to survive prolonged immersion and were later cast up on a beach and germinated, they would be unable to live in the beach environment. Islands smaller than $3\frac{1}{2}$ acres are effectively no more than beaches, because they are incapable of holding fresh water, and the flora is therefore restricted to species that are salt tolerant.[1] This leads to a corresponding reduction in the variety of animal life.

Variety of island habitats The inhospitability of the beach is merely an extreme example of the problem that faces any organism, even after it has succeeded in reaching an island. This is the problem of finding a habitat in which it can survive. It is obvious that a large island is more likely to contain a greater variety of habitats than a smaller one, and that it will therefore be able to support a greater variety of forms of life. For example, studies of a number of islands in the West Indies have shown that, when one island is approximately 10 times the size of

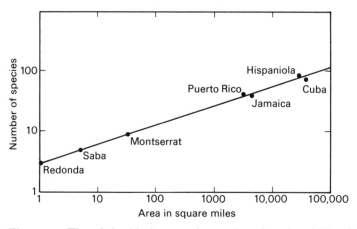

Figure 40. The relationship between the number of species of West Indian amphibians and reptiles and the area of the islands.

another, it contains approximately twice as many species of amphibians and reptiles (Fig. 40). Similarly, the diversity of available habitats increases in islands that include higher ground, providing a cooler, moister climate. Studies of the islands in the Gulf of Guinea, off West Africa, have shown that beyond a height of about 650 metres, they contain on average one additional native species of plant for every 3·5 metres of additional height.

It is difficult, however, to isolate the effects of any single factor such as area or height, since an island with hills or mountains has a larger surface area than one of the same size that is more or less flat. As appear, several other factors also affect the variety of an island's flora and fauna.

The very great importance of the diversity of habitat available can be shown by comparing the bat faunas of a number of islands off the northern coast of South America. Aruba, Curaçao, and Bonaire, which are arid, have few species, and only one more is found in Margarita, which has a little rain-forest. The numbers increase considerably in Grenada, Tobago, and Trinidad, where there are great areas of rain-forest, and especially in Trinidad, which has also some mountains.

Problems of isolation   In this last example, getting to the islands was not a great problem since we were considering bats, which can readily fly from island to island. In most cases, however, the biota (i.e. the fauna plus the flora) is strongly affected by the degree of isolation of the island. However diverse the habitats that it offers, the variety of the island life depends very much upon the rate at which colonizing animals and plants arrive. This, in turn, depends largely upon how far the island is from the source of its colonizers and upon the richness of that source. If the source is close, and if its fauna and flora are rich, then the island in its turn will have a richer fauna and flora than another, similar island which is more isolated or which depends upon a source with a more restricted variety of animals and plants. This is why the variety of organisms found on the Pacific islands becomes progressively poorer the farther they are from their main source—the Asian mainland. Each sea barrier further reduces the fauna and flora of the next island, which in turn becomes a poorer source for the next. The number of genera of cryptorhynchid weevils (a type of beetle) in the islands east of New Guinea shows this clearly (Fig. 41). Similarly, as can be seen from Figure 42, the diversity of Pacific island bird faunas does seem to be lower in islands which are isolated and small, and highest in those which are larger and close to their source (New Guinea).

Hazards of island life   Like any other population, the island population of a species must be able to survive periodic variations in its environment. But island life is more hazardous than that on the mainland, for several reasons. Catastrophe, such as volcanic eruption, has

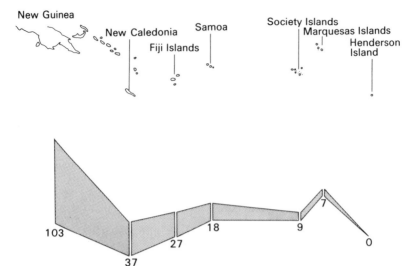

Figure 41. The number of genera of cryptorhynchid weevil progressively decreases in the islands East of New Guinea.

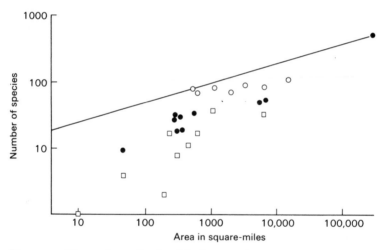

Figure 42. The numbers of land and fresh-water bird species on various islands and archipelagoes in the Pacific. The archipelagoes are widely scattered, and the distance effect is apparent in the great variance. Hawaii is included even though its fauna is derived mostly from the New World. "Near" islands (less than 500 miles from New Guinea) are shown as open circles, "far" islands (greater than 2,000 miles) are shown as squares, and islands at intermediate distances are shown as solid circles. A line is drawn through two of the islands nearest the source region and with the highest species densities in order to give a clearer idea of the degree of departure of the other islands from the potential densities of archipelagoes with high immigration rates. (After MacArthur and Wilson, 1963.)

longer-lasting effects in an island situation, for there is little opportunity for a species to vacate the area and return subsequently, neither is re-invasion easy should extinction take place. On the mainland, on the other hand, chance extinction of a species in a particular area can soon be made good by immigration from elsewhere.

Since its success and survival are the only measures of an organism's degree of adaptation to its environment, the fact that a species has become extinct also demonstrates that it was not well enough adapted. The adaptation to an island environment is an unusually difficult one for a species to make. In the first place, the immigrant individuals were originally part of the mainland population, and were therefore adapted to the mainland environment. They cannot, therefore, already be adapted to the different conditions of the island. Secondly, the colonists are usually few in number, and therefore can include only a very small part of the genetic variation that provided the mainland population with the flexibility to cope with environmental change. Finally, small populations are also far more susceptible to random non-adaptive changes in their genetic make-up. Since it is less likely to be closely adapted to its environment, a small population is also more liable to chance extinction.

A species which can make use of a wide variety of food is therefore at an advantage on an island, for its maximum possible population size will be greater than that of a species with more restricted food prefer-ences. The advantage of this will be especially great in a small island, in which the possible population sizes are in any case smaller. This is probably the reason why, for example, though on the larger islands of the Galapagos group, both the medium-sized finch *Geospiza fortis* and the small *G. fuliginosa* can co-exist, on some of the smaller islands of the group there is only a single form of intermediate size.[2]

Chance extinction is also a particular danger for the predators in the fauna, since their numbers must always be far lower than those of the species upon which they prey. As a result, island faunas tend to be unbalanced in their composition, containing fewer varieties of predator than a similar mainland area. This in turn reinforces the fundamental lack of variety of the animal and plant life of an island which is due to the hazards involved in entry and colonization. The complex inter-actions of continental communities containing a rich and varied fauna and flora act as a buffer that can cope with occasional fluctuations in the density of different species, and even with temporary local extinction of a species. This resilience is lacking in the simple island community, and so the chance extinction of one species may have serious effects and lead to the extinction of other species. All these factors increase the rate at which island species may become extinct.

The number of species found on an island therefore depends on a number of factors—not only its area and topography, its diversity of habitats, its accessibility from the source of its colonists, and the rich-ness of that source, but also the equilibrium between the rate of colonization by new species and the rate of extinction of existing species.

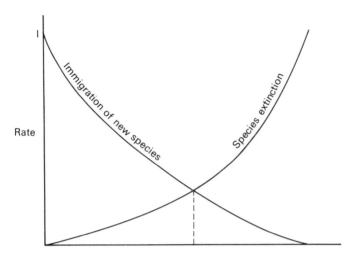

Number of species present

Figure 43. Equilibrium model of the biota of a single island. The equilibrial species number is reached at the intersection point between the curve of rate of immigration of new species, not already on the island, and the curve of extinction of species on the island. (After MacArthur and Wilson, 1963.)

MacArthur and Wilson[3] have in recent years provided a theoretical numerical model for the interaction of all these factors. Initially, they simply show how the position of the equilibrium point depends on the balance between the rates of colonization and of extinction (Fig. 43). The rate of colonization will be high initially, because the island will be reached quickly by those species which are adept at dispersal, and because these will all be new to the island. As time passes, more and more immigrants will find that they belong to species which have already colonized the island, so that the rate of appearance of new species will drop. The rate of extinction, on the other hand, will rise. This is partly because, since every species runs the risk of extinction, the more that have arrived the more species there are at risk. In addition, as more species arrive, the average population size of each will diminish as competition increases.

At first, the few species present can occupy a greater variety of ecological niches than would be possible on the mainland, where they are competing with many other species. If the island is now colonized by a new species which makes use of foodstuffs similar to those consumed by one of the earlier immigrants, competition between the two species will take place. As in the case of the closely related populations discussed in Chapter 4, this may result in the extinction of one of the two competitors, or in the gradual divergence of their food preferences so that the extent to which they are competing with one another is reduced. This latter process, the temporal or spatial separation of species described in Chapter 3, will mean that each is becoming more specialized in its requirements, making better use of a smaller variety of the possible sources of nourishment. For example, though three different insectivorous species of the Tanager (*Tanagra*) co-exist on the island of Trinidad, competition between them is reduced because they hunt for insects on different parts of the vegetation[4]: *T. guttata* searches mainly on the leaves, *T. gyrola* on the large twigs and *T. mexicana* mainly on the smaller twigs. If the variety of food used by each species is reduced in this way, it must follow that the size of the population of each species which the island can support is now smaller. Since the chances of extinction are greater for smaller populations, the rate of extinction must rise as new species colonize the island, until the equilibrium point is reached at which the rates of colonization and of extinction are equal.

In some situations the rate of extinction may not increase as rapidly as in MacArthur and Wilson's[3] model. For example, if the island initially contained no fauna or flora at all, there will be a gradual progression through a number of seral stages to the climax community. At each stage, some immigrants which would previously have been unable to establish themselves will now, for the first time, find a vacant niche. These changes may also, of course, lead to some extinctions among the earlier colonizing species.

So far, this model has considered rates only, and not the absolute numbers of species involved. As outlined above, the absolute numbers will depend upon the size of the island, and on its distance from the source of its colonists. MacArthur and Wilson show this, also, in graphic form (Fig. 44). They also give a mathematical analysis of the interaction of these factors which predicts surprisingly high rates of

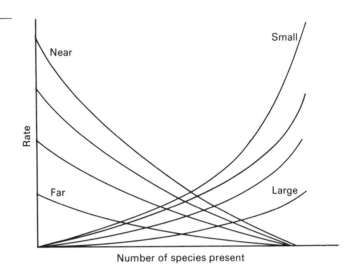

Figure 44. Equilibrium models of biotas of several islands of varying distances from the principal source area and of varying size. An increase in distance (near to far) lowers the immigration curve, while an increase in island area (small to large) lowers the extinction curve. (After MacArthur and Wilson, 1963.)

extinction. However, as they point out, such information as is available at present does support these estimates. For example, periodic studies have been made of the recolonization by birds of the East Indian islands of Krakatoa and Verlaten after the complete destruction of their fauna and flora in the volcanic explosion of 1883. By 1908, 13 species of bird had colonized Krakatoa. When further collections were made in 1919–21, the total of bird species had risen to 31, but two of the species found in 1908 were not collected or seen. By 1932–34, the total number of species (30) had remained almost stable, suggesting that this represents the equilibrium number for the island, but five species found previously were now absent. These figures suggest a rate of extinction of 0·2—0·4 species per year, i.e. that one species becomes extinct every 2–5 years—and the real figure may be higher than this, since it may be that a species could become extinct and reappear by recolonization between one collection period and the next. The figures are not very different from the extinction rate of 0·8–1·6 species per year predicted by MacArthur and Wilson's mathematical analysis. However, Williams'[5] studies on the ecology of colonization by the iguanid lizard *Anolis* in the West Indies suggests that extinction rates of terrestrial animals may be lower, since new waves of immigration, leading to competition and extinction, are far less frequent than in birds.

Opportunities for adaptive radiation Colonists encounter all these difficulties when they first enter an island, but there are rich opportunities for those species that can survive them long enough for evolution to adapt them to the new environment. These opportunities exist because of the lack of many of the parasites and predators that elsewhere would prey upon the species, and of many of the other species with which it normally competes. Like Darwin's finches on the Galapagos Islands, it may be able to radiate into ways of life not formerly available to it.

A good example of this can be found in the Dry Tortugas, the islands off the extreme end of the Florida Keys, which only a few species of ant have successfully colonized.[3] One species, *Paratrechina longicornis*, on the mainland normally nests only in open environments under, or in the shelter of, large objects; but on the Dry Tortugas, it also nests in environments such as tree trunks and open soil, which on the mainland are occupied by other species. Not every species, however, is capable of taking advantage of such opportunities in this way. The ant *Pseudomyrmex elongatus* is, on the Florida mainland, confined to nesting in red mangrove trees, occupying thin hollow twigs near the tree top. Though it has managed to colonize the Dry Tortugas, it is still confined there to this very limited nesting habitat.

Such opportunities for alterations in behavioural habits or in diet provide in turn the opportunity for the organism to become permanently adapted, by evolutionary change, to this new way of life. This process requires a longer period of time and is therefore unlikely to take place except on islands that are large and stable enough to ensure that the evolving species does not become extinct. But if an island does provide these conditions, then remarkable evolutionary changes may take place as colonizing species become modified to fill vacant niches.

One of the opportunities that may exist on an island often results from absence of one of the normal elements of a mainland flora—the tree. The seeds of trees are usually much larger and heavier than those of other plants and are therefore not readily transported long distances. As a result, other plants may develop to fill this vacant niche.[6] The modifications needed to produce a tree from a shrub which already possesses strong, woody stems are comparatively slight—merely a change from the many-stemmed, branching habit to concentration on a single, taller trunk. For example, though most members of the Rubiaceae are shrubs, this family has produced on Samoa the 8 metres tall tree *Sarcopygme*, which has a terminal palm-like crown of large leaves. Though more comprehensive changes are needed to produce a tree from a herb, many islands show examples of this phenomenon. In many cases, the plants involved are members of the Compositae, perhaps because they have unusually great powers of seed dispersal, are hardy and often already have partly woody stems. To this family belong both the lettuces, which have evolved into shrubs on many islands, and the sunflowers. On the isolated island of St. Helena in the South Atlantic can be found five different trees, 4–6 metres high, which have evolved in the island from four different types of immigrant sunflower (Fig. 45). Two of these (*Psiadia* and *Senecio*) are endemic species of more widely distributed genera, while the other three (*Commidendron*, *Melanodendron* and *Petrobium*) are recognized as completely new genera. These latter genera are therefore both endemic—that is, known only in St. Helena— and also autochthonous—that is, they actually evolved in the area concerned.

A similar process is responsible for the appearance on some islands of large animals, for example the Komodo dragon (*Varanus komodoensis*), a giant lizard which lives on Komodo Island and nearby Flores Island in the East Indies. These animals increased in size to occupy niches which on the mainland are filled by animals much larger than typical varanid lizards.

Dwarfs and flightless birds All these organisms evolved in islands to fill habitats normally closed to them. But other evolutionary changes frequently found on islands are the direct result of the island environment itself, not of the restricted fauna and flora. We have seen how serious may be the effect of a small population. But the same island will be able to support a larger population of the same animal if the size of each individual is reduced. This evolutionary tendency on islands is shown by the find of fossil pygmy elephants that once lived on islands in both the Mediterranean and the East Indies. On a much smaller scale, the size of lizards on four of the Canary Islands still shows the same

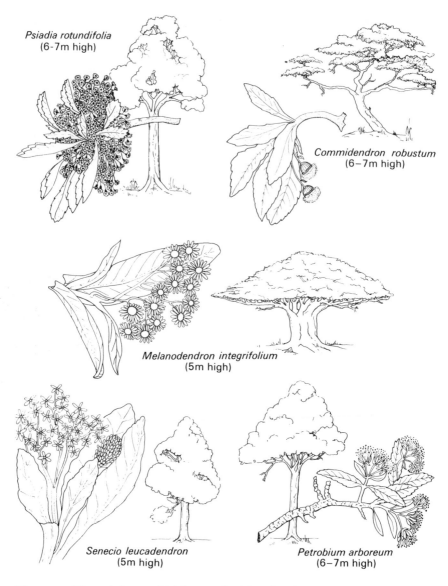

Figure 45. The varied trees which have evolved from immigrant sunflowers on St. Helena Island.

phenomenon. The head to vent length of males of the lizard *Lacerta galloti* ranges from 135 mm on the largest island, Tenerife, to only 82 mm on Hierro, the smallest.

Another tendency is for island species to lose the very dispersal mechanisms that allowed them to reach their home. Once on the restricted area of the island, the ability for long-distance dispersal is no longer of value to the species: in fact it is a disadvantage. The seeds of plants tend to lose their "wings" or feathery tufts, and many island insects are wingless. The loss of wings by some island birds may be partly for this reason, and partly because there are often no predators from which to escape. A few out of many examples are the kiwi and moa of New Zealand, the elephant birds of Madagascar, and the dodo of Mauritius (the last three are extinct, but only because man was the predator).

The Hawaiian Islands As has been seen, there are many aspects of island life that are unique, and many others that differ only in degree from life on the continental land masses. The result of the action of all these different factors can be seen by examining the flora and fauna of one particular group of islands. The Hawaiian Islands provide an excellent example, for they form an isolated chain, 2650 km long, lying in the middle of the North Pacific, just inside the tropics (Fig. 46). The islands are of volcanic origin, rising steeply from a seafloor which

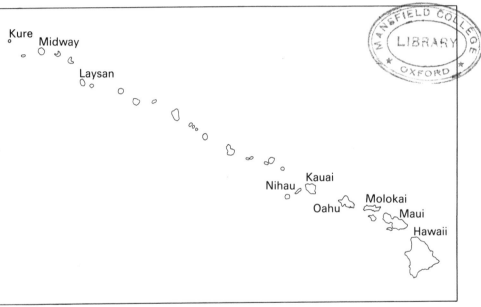

Figure 46. Hawaiian island chain.

is 5500 m deep to the volcanic peaks which reach up to 4250 m. Hawaii itself lies 3200 km from North America and 5500 km from Japan. The islands seem to be the result of the activity of a particular point in the Earth's interior, past which the sea floor has been moving westwards. The most westerly island at the end of the long Hawaiian chain, Kure, is therefore the oldest (about 15 million years), while the most easterly, Hawaii, is the youngest (less than 700,000 years), with the still active volcano Kilauea.

The most obvious result of the extreme isolation of the Hawaiian Islands is that many groups are completely absent. There are no truly freshwater fish and no native amphibians, reptiles, or mammals (except for one species of bat). The difficulty of getting to the islands is shown by the fact that the present-day bird fauna of Hawaii seems to be the result of only 15 different colonizations.

Most of the birds are of American origin, but the closest relatives of most of the Hawaiian animals and plants live in the Indo-Malayan region. For example, of the 1729 species and varieties of Hawaiian seed plants, 40 per cent are of Indo-Malayan origin but only 18 per cent are of American origin; also, nearly half of the 168 species of Hawaiian ferns have Indo-Malayan relatives, but only 12 per cent have American affinities. This is not surprising, for the area to the south and west of the Hawaiian chain contains many islands, which can act as intermediary homes for migrants, while that to the east is almost completely empty.

The way in which the Hawaiian birds reached the islands is obvious enough. One of the plants which probably came with them is *Bidens*, a member of the Compositae, whose seeds are barbed and readily attach themselves to feathers. The Hawaiian insects, too, arrived by air. Entomologists have used aeroplanes and ships to trail fine nets over the Pacific at different heights and have trapped a variety of insects, most of which, as would be expected, were species with light bodies. These types also predominate in the Hawaiian islands—an indication of their airborne arrival—though heavier dragon-flies, sphinx-moths, and butterflies are also found there.

Many other plants must originally have arrived as wind-borne seeds. One that is particularly interesting is a tree, *Metrosideros*. It is unusual because its seeds are tiny compared with those of other trees, and this has allowed it to become widely dispersed throughout the Pacific islands —it is now the dominant tree of the Hawaiian forests. It is a pioneering tree, able to form forests on virtually soil-less lowland lava-rubble—a great advantage on a volcanic island. It is, however, a tolerant species, and is able to grow as a small shrub in bogs at high altitudes.

Apart from the ubiquitous coconut, another plant found in Hawaii is *Scaevola sericea*, which was able to cross the sea barrier with its white buoyant fruits; this shrub forms dense hedges along the edge of the beach on Kauai Island. Another sea-borne migrant is *Erythrina*; most species of this plant genus have buoyant bean-like seeds. On Hawaii, after its arrival on the beach, *Erythrina* was unusual in adapting to an inland environment, and a new endemic species, the coral tree *Erythrina sandwichensis*, has arisen on the island. Unlike those of its ancestors, the seeds of the coral tree do not float—an example of the loss of its dispersal mechanism often characteristic of an island species.

The successful colonists of the Hawaiian islands are the exceptions; many groups have failed to reach them. As already noted, no terrestrial vertebrates occur there naturally, while 21 orders of insect are completely absent. As might be expected, most of these are types that seem in general to have very limited powers of dispersal. For example, the Formicidae (ants and termites), which are an important part of the insect fauna in other tropical parts of the world, were originally absent for this reason. They have, however, since been introduced by man, and 36 different species have now established themselves and filled their usual dominant role in the insect fauna. This proves that the obstacle was entrance to the islands, not the nature of the Hawaiian environment.

As ever, the absence of some groups has provided greater opportunities for the successful colonists. Several insect families, such as the crickets, fruit-flies, and carabid beetles, are represented by an extremely diverse adaptive radiation of species, each radiation derived from only a few original immigrant stocks.[7] The fruit flies, belonging to the closely-related genera *Drosophila* and *Scaptomyza*, have been studied in particular detail.[8] Of all the known species (over 1200), about one-third are found only in the Hawaiian chain. The fact that the island of Hawaii itself is less than 700,000 years old also means that the species on that island must have differentiated within that time. Detailed studies of the chromosome structure of the Hawaiian fruit flies are now even making it possible to reconstruct the sequence of colonizations which must have taken place.[9]

The abundance of species of fruit flies in these islands is probably due partly to the great variations in climate and vegetation to be found there, and also to the periodic isolation of small islands of vegetation by lava flows. But another major factor has been that the Hawaiian fruit flies, in the absence of the normal inhabitants of the niche, have been able to use the decaying parts of native plants as a site in which their larvae feed and grow. This change is probably also due to the fact that their normal food of yeast-rich fermenting materials is rare in the Hawaiian islands.

The same phenomenon of a great adaptive radiation has taken place in other groups of animals and plants. In general, therefore, though the islands contain comparatively few different families, each contains an unusual variety of species, nearly all of which are unique to the islands. In fact, out of the whole of the Hawaiian flora, over 90 per cent of the species are endemic to the islands.

There are many other examples of Hawaiian adaptive radiations, but three are of particular interest: the tarweeds and the lobeliads among the plants,[6] and the honey-creepers among the birds. The tarweeds belong to the family Compositae, and probably arrived on the islands as sticky seeds attached to the feathers of birds. They have produced only three genera in the islands (*Dubautia, Argyroxiphium*, and *Wilkesia*), but these have colonized a variety of habitats. For example, on the bare cinders and lava of the 3050 m high peak of Mt. Haleakala on the island of Maui, two of the few plant species which can survive are tarweeds. *Dubautia menziesii* is adapted to this arid environment by its tall stem and stubby succulent leaves, while the silversword *Argyroxiphium sandwichense* is covered by silvery hairs that reflect the heat. A few hundred metres below the bare volcanic peaks, conditions are at the other extreme, because most of the rain falls at heights of from 900 to 1800 m; these regions receive from 250 to 750 cm of rain per year. The upper regions of 1770 m high Mt. Puu Kukiu on Maui are covered by bog in which thrives another silversword, *A. caliginii*. On the island of Kauai the heavy rainfall has led to the development of dense rain forest, in which *Dubautia* has evolved a tree-like species, *D. knudsenii*, with a 0·3 m-thick trunk and large leaves to gather the maximum of sunlight in the dim forest. Kauai bears another tarweed which shows the tendency for island plants to become trees. In the drier parts of this island grows *Wilkesia gymnoxiphium* with a long stem which carries it above the shrubs that compete with it for light and living space. This species also shows another example of the loss of the dispersal mechanism that first brought the ancestral stock to the island: the seeds of *Wilkesia* are heavy and lack the fluffy parachutes usually found among the Compositae.

Lobeliads (members of the plant family Campanulaceae) are found in all parts of the world, but they have undergone an unusual adaptive radiation in the Hawaiian Islands, because their normal competitors, the orchids, are rare. The Hawaiian lobeliads include 150 endemic species and varieties, making six endemic genera. Over 60 species of one endemic genus alone (*Cyanea*) are known, showing an incredible diversity of leaf form (Fig. 47). The plants range from the 9 m-tall tree *C. leptostegia* (similar in appearance to the tarweed *Wilkesia*) to the 0·9 m-tall, soft-stemmed *C. atra*. The species of another genus,

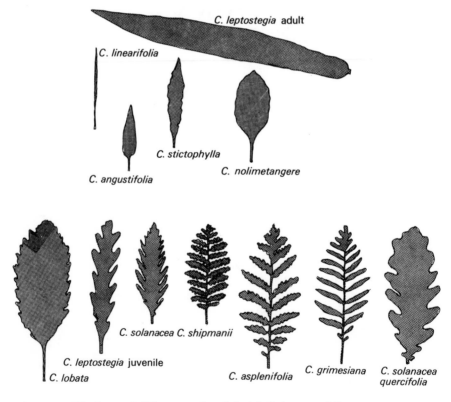

Figure 47. The leaves of different species of the lobeliad genus of *Cyanea*.

Clermontia, are less varied in overall size, but are very varied in the size, shape, and colour of their flowers. These are mainly tubular and brightly coloured, a type of flower which is often associated with pollination by birds. On isolated islands such as Hawaii, the adaptation of larger flowers to pollination by birds may be because the large insects that would normally pollinate such flowers on the mainland are absent. It is no coincidence that the adaptive radiation of the Hawaiian lobeliads has been accompanied by the adaptive radiation of a nectar-eating type of bird, the honey-creepers.[10]

The ancestor of these birds was probably a finch-like immigrant from America which fed on insects and nectar. From the original immigrants adaptive radiation has produced 9 endemic genera, comprising the endemic family Drepanididae (Fig. 48). Many of the genera, such as *Himatione, Vestiaria, Palmeria, Drepanidis,* many species of *Loxops,* and one species of *Hemignathus (H. procerus)* are still nectar eaters, feeding from the flowers of the tree *Metrosideros* and the lobeliad

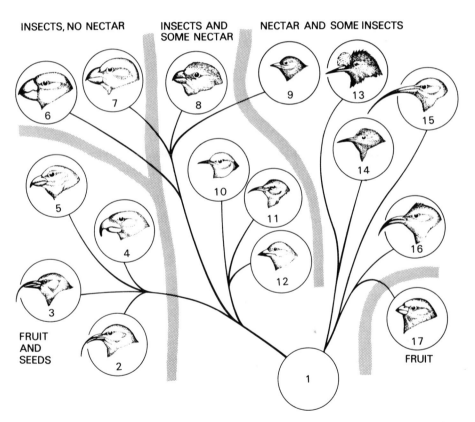

INSECTS, NO NECTAR   INSECTS AND SOME NECTAR   NECTAR AND SOME INSECTS

FRUIT AND SEEDS   FRUIT

Figure 48. The evolution of dietary adaptations in the beaks of Hawaiian honeycreepers. 1, Unknown finch-like colonist from North America. 2–5, *Psittacirostra psittacea, P. kona, P. bailleui, P. cantans.* 6, *Pseudonestor xanthophrys.* 7–9, *Hemignathus wilsoni, H. lucidus, H. procerus.* 10–12, *Loxops parva, L. virens, L. coccinea.* 13, *Drepanidis pacifica.* 14, *Vestiaria coccinea.* 15, *Himatione sanguinea.* 16, *Palmeria dolei.* 17, *Ciridops anna.*

Clermontia. Since insects, too, are attracted to the nectar, it is not surprising to find that many nectar-eating birds are also insect-eaters, and from this it is a short step to a diet of nothing but insects. *Hemigna-thus wilsoni* uses its mandible, which is slightly shorter than the upper half of its bill, to probe into crevices in bark for insects, and *Pseudonestor kona* uses its heavier bill to rip open twigs and branches in search of insects. Species of the genus *Psittacirostra* too, have heavy, powerful beaks, which some use for cracking open seeds, nuts or beans. The lighter bill of the recently extinct *Ciridops* was used for eating the soft flesh of the fruits of the Hawaiian palm *Pritchardia*.

Living as they do on fruits, seeds, nectar, and insects, it is not surprising that none of the drepanidids show the island-fauna charac-teristic of loss of flight. However, both on Hawaii and on Laysan to the west, some genera of the Rallidae (rails) have become flightless (a common occurrence in this particular family of birds). The phenomenon of flightlessness is also common in Hawaiian insects: of the endemic species of carabid beetle, 184 are flightless and only 20 fully winged. The Neuroptera or lacewings are another example—their wings, usually large and translucent, are reduced in size in some species, while in other species they have become thickened and spiny.

To summarize, islands provide a unique opportunity to study evolution, for their small, impoverished faunas and floras are the ideal situation for rapid evolutionary modification and adaptive radiation. At the same time, island life is unusually hazardous, so that there is a complex interaction between the processes of immigration, colonization and extinction, and quantitative analysis of this has only recently commenced.

References

1 WHITEHEAD D.R. & JONES C.E. (1969) Small islands and the equilibrium theory of insular biogeography. *Evolution* **23,** 171–179.
2 LACK D. (1969) Subspecies and sympatry in Darwin's finches. *Evolution* **23,** 252–263.
3 MACARTHUR R.H. & WILSON E.O. (1967) *The Theory of Island Biogeography.* Princeton University Press, Princeton.
4 SNOW B.K. & SNOW D.W. (1971) The feeding ecology of tanagers and honeycreepers in Trinidad. *Auk* **88,** 291–322.
5 WILLIAMS E.E. (1969) The ecology of colonization as seen in the zoo-geography of anoline lizards on small islands. *Quart. Rev. Biol.* **44,** 345–389.
6 CARLQUIST S. (1965) *Island Life.* Natural History Press, New York.
7 ZIMMERMAN E.C. (1948) *The Insects of Hawaii*: 1, *Introduction.* University of Hawaii Press, Honolulu.
8 ROBERTSON F.W. (1970) Evolutionary divergence in Hawaiian *Drosophila. Sci. Prog., Oxf.* **58,** 525–538.
9 CARSON H.L. (1970) Chromosome tracers of the origin of species. *Science* **168,** 1414–1418.
10 AMADON D. (1950) The Hawaiian honey-creepers. *Bull. Am. Mus. nat. Hist.* **95,** 151–262.

CHAPTER 6

THE DISTANT PAST

As explained in Chapter 2, one way of grouping the terrestrial biological communities of the world is to place each of them in one of eight biomes. Each of these is distinguished from the others by its characteristic climate. A particular biome, such as desert, may therefore exist in many different parts of the world. In each desert live animals and plants of broadly similar appearance and way of life, but these may belong to quite different groups from those found in a similar desert in another part of the world. A map showing the distribution of the biomes therefore tells us nothing about the patterns of distribution of taxonomic groups of animals or about the way in which different groups replace one another.

An alternative approach to classifying the patterns of distribution is to subdivide the world's surfaces into regions which appear to differ from one another in the dominant types of plant or animal to be found there. Although very few groups have precisely the same pattern of geographical distribution, there are some zones which mark the limits of distribution of many groups. This is because these zones are barrier regions, where conditions are so inhospitable to most organisms that few of them can live there. For terrestrial animals, any stretch of sea or ocean proves to be a barrier of this kind—except for flying animals whose distribution is for this reason obviously wider than that of solely terrestrial forms. Extremes of temperature, such as exist in deserts or in high mountains, constitute similar (though less effective) barriers to the spread of plants and animals.

These three types of barrier—oceans, mountain chains and large deserts—therefore provide the major discontinuities in the patterns of the spread of organisms around the world. Oceans completely surround Australia. They also virtually isolate South America and North America from each other and completely separate them from other continents. Seas, and the extensive deserts of North Africa and the Middle East, effectively isolate Africa from Eurasia. India and south-east Asia are similarly isolated from the rest of Asia by the vast, high Tibetan Plateau, of which the Himalayas are the southern fringe, together with the Asian deserts which lie to the north.

Each of these great land areas, together with any nearby islands to which its fauna or flora has been able to spread, is therefore comparatively isolated. It is not surprising to find that the patterns of distribution of both the faunas (*faunal provinces* or *zoogeographical regions*) and the floras (*floral realms*) largely reflect this pattern of geographical barriers. The two schemes differ only in that the pattern of floral realms shows a closer relationship to the latitudinally-determined pattern of climate. Thus the north temperate areas (North America and Eurasia) are linked together into a single floral realm, as are the Old World tropical areas (Africa and India), while the temperate southern extremities of South America and of Africa are recognized as separate floral realms.

Before the detailed composition of these faunal provinces and floral realms can be understood fully, it is first necessary to explain the ways in which today's patterns of geography, climate and distribution of life came into existence. From what has been discussed in earlier chapters, it is clear that the differences between the faunas and floras of different areas might be due to a number of factors. Firstly, any new group of organisms will appear first in one particular area. If it competes with another, previously established group in that area, the expansion in the range of distribution of the new group may be accompanied by contraction in that of the old. However, once it has spread to the limits of its province or realm, whether or not it is able to spread into the next will depend initially on whether it is able to surmount the geographical ocean or mountain barrier, or to adapt to the different climatic conditions to be found there. (Though even if it is able to cross to the next province or realm, it may be unable to establish itself because of the presence there of another group which is better adapted to that particular environment.) Of course, changes in the climatic or geographical pattern could lead to changes in the patterns of distribution of life. For example, gradual climatic changes, affecting the whole world, could cause the gradual northward or southward migrations of floras and faunas, because these extended into newly favourable areas and died out in areas where the climate was no longer hospitable. Similarly, the possibilities of migration between different areas could change if vital links between them became broken by the appearance of new barriers, or if new links appeared.

Until recently, it seemed as though the geographical ranges of different animals and plants could be explained using only three basic principles—evolution, climatic change and land bridges. Such a belief was supported by evidence from the most recent (and therefore best documented) past. The Pleistocene changes in the ranges of the plants and animals of North America and Eurasia could be straightforwardly explained as the results of expansions and contractions of the ice-sheets which covered the northern parts of the two continents during much of this period. These recent changes, which are described in the next chapter, presumably resulted from more general changes in the Earth's climate. In North America, studies had been made of the thick sediments, eroded from the Rocky Mountains, that had been deposited over the mid-western part of the country. Extending back for the whole of the Tertiary Period, which (as can be seen from the geological time scale, Fig. 49) lasted nearly 65 million years, these too showed that the climate of the North American continent had gradually become colder during that time. The effects of this cooling had included the Bering

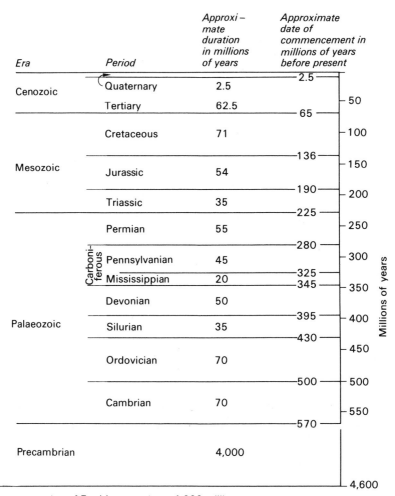

Era	Period	Approximate duration in millions of years	Approximate date of commencement in millions of years before present	Millions of years
Cenozoic	Quaternary	2.5	2.5	50
	Tertiary	62.5	65	
Mesozoic	Cretaceous	71	136	100 / 150
	Jurassic	54	190	200
	Triassic	35	225	
Palaeozoic	Permian	55	280	250
	Pennsylvanian (Carboniferous)	45	325	300
	Mississippian (Carboniferous)	20	345	350
	Devonian	50	395	400
	Silurian	35	430	
	Ordovician	70	500	450 / 500
	Cambrian	70	570	550
	Precambrian	4,000		

Formation of Earth's crust about 4,600 million years ago

Ages of the Epochs of the Cenozoic Era
Pleistocene Epoch	2.5— 0	million years ago
Pliocene Epoch	5.0— 2.5	" " "
Miocene Epoch	22.5— 5.0	" " "
Oligocene Epoch	37.5—22.5	" " "
Eocene Epoch	53.5—37.5	" " "
Paleocene Epoch	65.0—53.5	" " "

Figure 49. Geologic time scale.

area between Siberia and Alaska which, as a result, had gradually become inaccessible as a migration route to all but the most hardy animals and plants.

Though it was clear that such geological events as the rising of mountain chains must inevitably have affected the climate of surrounding areas, there seemed until recently no reason to search for new principles to explain the distribution of animals, plants or climates during the whole of the Tertiary, and the same could also be said for much of the preceding Mesozoic Era. But studies of the late Palaeozoic and early Mesozoic Eras (about 225–300 million years ago) provided evidence of a very different world.

Gondwanaland and Laurasia   One of the first clues to this was the discovery that the flora that had covered the southern continents during the Permian was rather different from that found in the northern continents. One of the groups common to all the floras was the seed-ferns, which, as their name implies, were very like living ferns but bore large seeds instead of small spores. The Permian southern flora was dominated by seed-ferns belonging to the genera *Glossopteris* and *Gangamopteris*—it is usually called the *Glossopteris* flora. Its distribution extended over South America, South and Central Africa, Australia, Antarctica, and India. All these areas are collectively known as *Gondwanaland*; North America, Greenland, and Eurasia, from which the *Glossopteris* flora was absent, are known as *Laurasia*. The presence of this flora in India was unexpected, because on the basis of present-day geography one might expect its flora to be like that of Asia and unlike that of the southern continents. India was also found to be peculiar in another way. Shortly before the appearance of the *Glossopteris* flora, all the southern continents had been affected by a giant glaciation. The evidence for this lies in the thick drift deposits of rock-rubble or *tillite* that these glaciers had carried, and that had been left behind as they melted. Surprisingly, the southern glaciation had affected not only Australia, South America, and South Africa, but also India. Furthermore, the scratches that they made on the underlying rocks show that the glaciers invaded India from the south—that is, from the direction of the present Equator.

The idea of continental drift   The unexpected distributions of Permian and Carboniferous floras and glaciers could most simply be explained if the continents had at that time been joined together in a pattern different from that seen today, and had also lain farther south than they do now (Fig. 50). They would then have had to drift apart and northward to reach their present positions. Several scientists in the early years of the 20th century had noticed the great similarity between the outlines of the continents on either side of the Atlantic, and had suggested that they had once been joined and later moved apart. In 1910 the American geologist F. B. Taylor, for example, suggested that movements of this kind might have been responsible for the formation of the major mountain ranges of the world. Up to that time, however, the possibility of continental movement was put forward as an explana-

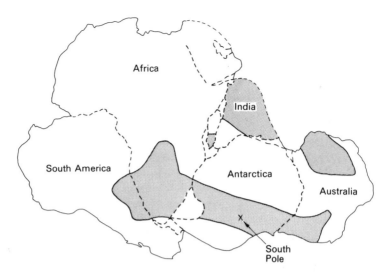

Figure 50. The Gondwanaland continents (plus Madagascar and New Zealand) probably fitted together in this pattern—the edges of the continental shelves, not the coastlines, are shown. Glaciers are known to have been present in the shaded areas during the Late Carboniferous and Early Permian, and they must also have covered much of the intervening regions. The approximate position of the South Pole at that time is shown.

tion of isolated phenomena of geology, without any real attempt to analyse all the different results such movements might have had, or to find evidence that they *had* happened. Such analyses and evidence were first provided by the German scientist Alfred Wegener. He, too, had been impressed by the similar shapes of the coastlines on either side of the Atlantic, but in the autumn of 1911 he accidentally chanced upon a report containing details of palaeontological evidence for an ancient land connection between Brazil and Africa. Further research into the literature of palaeontology and geology soon provided a wealth of additional evidence. For example, Wegener found that, if the edges of the continental shelves on either side of the Atlantic were fitted together, many geological features that had previously ended abruptly at the edge of one continent were now continuous with a similar feature on the adjacent continent. At the same time, his reconstruction solved the problems of the extent of the great southern glaciation and of the distribution of the *Glossopteris* flora.

A geological controversy Wegener's theory, which was first published in 1915, demanded radical changes in the beliefs of scientists in many fields of geology and related sciences. For this reason alone, it would probably have been slow to find acceptance even if it had come from a worker with a long history of research and publication in geology. Wegener, however, was primarily a meteorologist and astronomer, whose earlier published work had been on the thermodynamics of

113

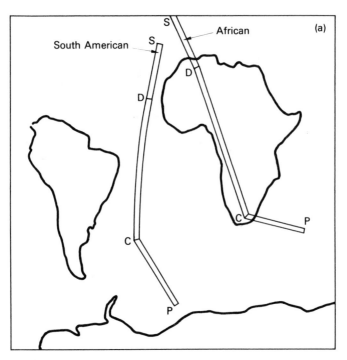

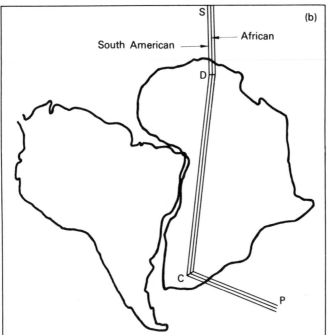

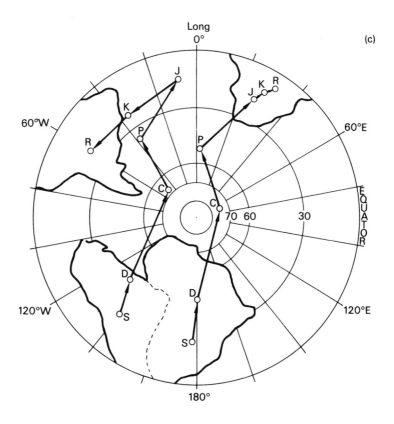

Figure 51. Polar wandering patterns of South America and Africa. (a) With the two continents in their present positions. The position of the South Magnetic Pole relative to each continent is shown for the Silurian (S), Devonian (D), Carboniferous (C) and Permian (P). (b) The continents can be moved so that their polar wandering patterns overlap, proving that they moved in unison during this period of time (note that the pattern of fitting of the continents is the same as that produced by matching the edges of their continental shelves). (c) The paleomagnetic data suggests that these two continents moved across the South Magnetic Pole as shown here. The positions of a particular locality in each continent are shown in the Silurian (S), Devonian (D), Carboniferous (C), Permian (P), Triassic-Jurassic (J), Cretaceous (K) and today (R).

the atmosphere. Geologists were, perhaps naturally, even less ready to change the whole basis of their approach to the structure of the Earth at the suggestion of such an outsider. Many scientists "solved" the problem to their own satisfaction by suggesting solutions that left the theories in their own branch of science unchanged at the expense of radical changes in other sciences. For example, many biologists suggested that the different parts of Gondwanaland had been connected by land bridges that had allowed animals and plants to migrate between them. Geologists at that time did not know enough about the ocean beds to prove that such land bridges were impossible. Some geologists even suggested that the glaciers had never existed, the supposed glacial tillites being the result of mud flows of some kind. Others accepted the tillites, but thought that their widespread distribution must have resulted from a world-wide change to a very cold climate—though there was no evidence of contemporary glaciation in Laurasia. But the main objection to Wegener's theory was that no forces were known that were strong enough to move whole continents. Only since about 1953, when a variety of new lines of evidence emerged, have Wegener's ideas gradually become accepted.[2]

The evidence of palaeomagnetism Study of the direction of magnetization of rocks has provided one of these new supports for the occurrence of continental drift. Some rocks contain magnetized particles that, as the rocks were being deposited or (in the case of lavas) were cooling down, became aligned along the lines of the Earth's magnetic field. These rocks therefore contain, as it were, many tiny compass needles that still tell us how far away, and in which direction, the magnetic poles lay when the rocks were formed. Obviously, if the continents had never moved, these fossil compasses should all point to the present magnetic poles—but they do not. Instead, if a series of rocks of different ages from within one continent are studied, and the positions of one of the magnetic poles at these different times are plotted on a map, it looks as though the pole has gradually moved across the Earth's surface. We know that the magnetic poles do wander a little, but it seems unlikely that they have ever lain far from the geographical poles. This apparent movement is therefore more easily explained as the result of movements of the continents themselves (Fig. 51).

These *palaeomagnetic* measurements not only show that the continents have drifted relative to the magnetic poles, but also prove that they have drifted relative to one another. If they had not, then all the fossil compasses of a single age, from all the different continents, should point to the same place—but they do not. For the past 70 million years or so, the paths of movement of the different continents are quite unlike one another. But, if we take a globe and move the continents back along their old paths, we find that they gradually come together in a pattern very like that which Wegener suggested. At that point in time, therefore, the continents first started to separate. Before that time they had all moved in unison as part of the two connected supercontinents, Laurasia and Gondwanaland.

Break-up of the supercontinents Not surprisingly, our knowledge of the precise relationships between the different continents becomes progressively less certain as one goes back in time. Some evidence suggests that there were three separate land masses in the Lower Palaeozoic —Gondwanaland, North America/Europe, and Asia— and that it was the union of these blocks that raised the Appalachian, Caledonian and Ural mountain ranges, and formed the single land mass that is often called *Pangaea*. Whatever its earlier history may have been, this single land mass seems to have altered greatly in its geographical position during the Palaeozoic and early Mesozoic (Fig. 52). About 420 million years ago, North America and Africa lay close together around the South Pole. Gondwanaland therefore lay on the far side of the South Pole, so that what now appear as the southern ends of its continents pointed up toward the Equator. The land masses moved northward in such a way that nearly the whole length of Africa gradually passed across the South Pole. Eventually, in the late Carboniferous to early Permian periods, about 270 million years ago, Antarctica and Australia lay in the region of the South Pole while Europe lay on the Equator.

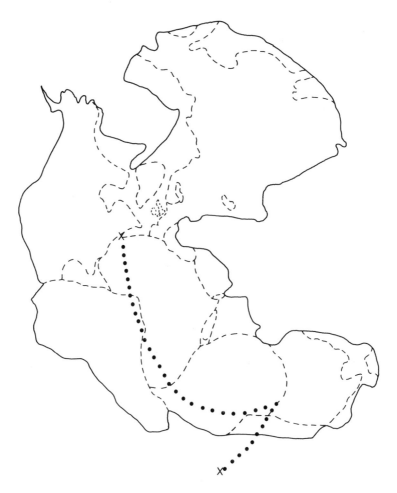

Figure 52. Pangaea as it existed in the Permian and Triassic (since its position is uncertain, S.E. Asia is omitted). Dotted lines indicate present-day continental coastlines. The successive positions of the South Magnetic Pole from the Cambrian to the Jurassic are indicated by a series of circles.

This was the time of the great Gondwanaland glaciations we have already mentioned, and of the tropical coal forests of Europe and North America.

So far, Pangaea had remained whole. Now, while continuing to move northward, it began to disintegrate. The history of this break-up into separate continents, and their movements, are shown in Figs. 53–55. The true edges of the land masses are not marked by their coastlines but by the edge of the continental shelf. In the past, as at

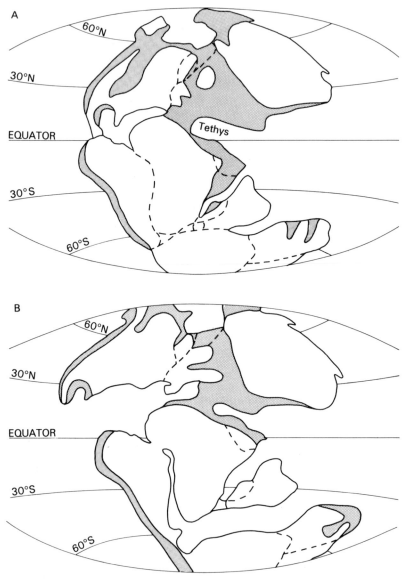

Figure 53. The world in A, the Late Jurassic, 140 million years ago; B, the Mid Cretaceous, 105 million years ago. Dotted lines indicate the present-day continental coastlines, shaded areas indicate shallow epicontinental seas.

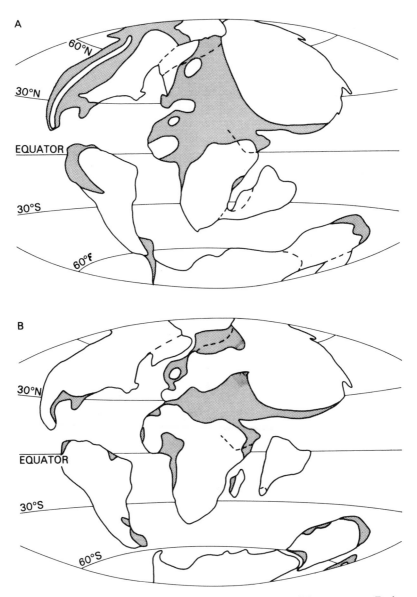

Figure 54. The world in A, the Late Cretaceous, 75 million years ago; B, the Lower Tertiary (Upper Eocene), 50 million years ago. Dotted lines indicate present-day continental coastlines, shaded areas indicate shallow epicontinental seas.

present, shallow "epicontinental" seas often covered the fringes of the land masses, like the North Sea at the edge of Europe today, or penetrated deeper into the continent, like Hudson's Bay today. Though the extent of these shallow seas varied, they must have formed a barrier to the spread of terrestrial organisms, and they are shown in Figs. 53–55. They were particularly extensive in the Jurassic and Cretaceous.

A

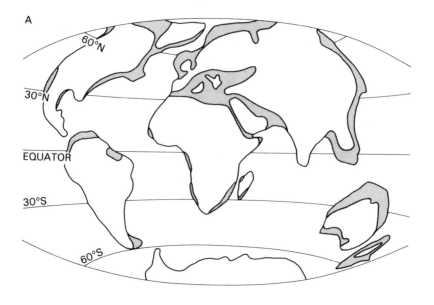

B

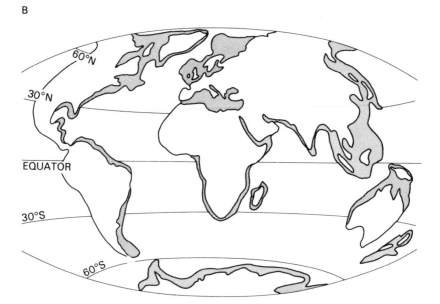

Figure 55. The world in A, the Upper Tertiary (Miocene), 20 million years ago; B, today. Shaded areas indicate shallow epicontinental seas.

Though an ocean known to geologists as the Tethys Ocean had long formed a great embayment in the easterly margin of Pangaea, the first break in the single land mass took place in the Late Jurassic (Fig. 53A). At this time a slight movement apart of North America and Africa formed the first, narrow strip of Atlantic—though North America remained attached to Europe, and Europe to Africa. As Africa continued to swing away from North America, it later started to separate from South America, a split forming around the southern end of Africa in the Mid Cretaceous (Fig. 53B).

As Africa swung yet further away from North America, Arabia gradually came closer to southern Asia, narrowing the Tethys Ocean till the two land masses met in the Late Cretaceous (Fig. 54A). South America was now quite separate from Africa. The North Atlantic, between North America and Europe, also began to open at this time, but the break between the two continents was not complete until the separation of Greenland from Europe in the Lower Tertiary (Fig. 54B). It was also at this time that Australia first split from Antarctica and began to move northwards. Finally, it was not until the Upper Tertiary (Fig. 55A) that India collided with Asia, throwing up the Himalayan Mountains, and a smaller northward movement of Africa into Europe caused the mountain chains of the Alps and Carpathians in southern Europe. As far as we know, all these slow movements are still continuing. The last of today's links to form was the Panama Isthmus between North and South America. (Though it was probably originally part of Gondwanaland, the history of South East Asia is uncertain, and it is not shown on the maps before the Miocene.)

The driving force Not only is much more now known about the detailed movements of the continents, but the force that makes them move may have been identified. The interior of the Earth is hotter than its surface. As a result, there may be great convection currents bringing hot material from the deeper layers to the surface and these may be the cause of the movements of the continents (Fig. 56). For example, such a current reached the surface forming a great ridge that runs down the middle of the South Atlantic. This line originally lay *within* the super-continent of Gondwanaland, but the spreading convection currents gradually separated the land masses on either side of the line—South America and Africa. The new material, brought up from within the Earth and spread on either side of what is now called the mid-Atlantic Ridge, formed the floor of the new Atlantic Ocean. Similar processes led to the fragmentation of the rest of Gondwanaland, and to the separation of North America, Greenland and Eurasia.

The spreading apart of all these areas, and the formation of new seas between them, is accompanied by the disappearance of old regions of sea floor. This takes place at the oceanic trenches, where the sea floor disappears downwards into the Earth. One result of this process is that none of the deep sea floor is more than about 150 million years old, whereas continental rocks as old as 3500 million years are known.

Effect on climate The movements of the continents appear to have been quite slow—only about 5 to 10 cm a year. These movements must have affected life in several ways, even through the changes must have been incredibly gradual, and noticeable only over a period of millions

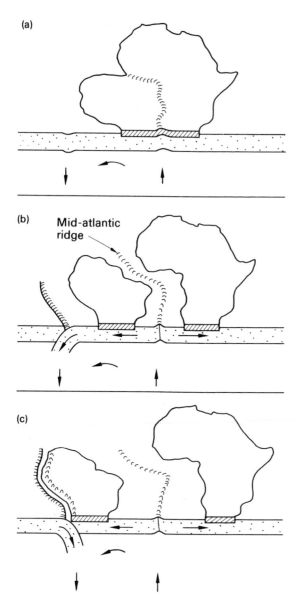

Figure 56. How South America and Africa drifted apart. (a) The two continents were originally part of a single super-continent. A convection cell appeared in the deeper layers of the earth, an upward continent appearing under the layers of the super-continent, and a downward current from the west. (b) The super-continent splits above the ascending current. New coastal material (flooring the new South Atlantic Ocean) appears on either side of this split line, which now forms the mid-Atlantic ridge. At the same time, an ocean trench forms above the descending current and old coastal material disappears into this. (c) South America has moved westward until it is adjacent to the oceanic trench. The coastal material consumed by the trench is now derived from the west; its movement below western South America causes earthquakes, and the raising of this region forms the Andes.

of years. The most obvious change would have resulted directly from the movement of the supercontinents relative to the poles of the Earth, and relative to the Equator. As they moved, so the different areas of land would eventually have come to lie in cold polar regions, in the hot, wet equatorial regions, in the dry sub-tropical regions, or in the cool, damp temperate regions. Drift also affected the climates of the continents in two less direct ways. Before the break-up, much of the area of the supercontinents lay far from the sea and must have had fairly stable climates. More variable, wetter weather must have affected greater areas when the seas gradually spread between the drifting continents as they separated.

Furthermore, the distribution of climate within the continental masses must also have been affected by the appearance of new mountain ranges as a result of continental drift. These would have had particularly great effects on the climate of the continents if they arose across the paths of the prevailing moisture-bearing winds, since areas in the lee of the mountains would then become desert. These can be seen today in the Andes, to the east of the mountain chain in southern Argentina, and to the west along the coast from northern Argentina to Peru; the winds in these two regions blow in opposite directions. A huge area of desert, including the arid wastes of the Gobi Desert of outer Mongolia, has also formed in central Asia, far from seas from which winds could gain moisture.

Life in the moving continents Neither the details of the precise timing of the break-up of Laurasia and Gondwanaland, nor the complete fossil record of animals and plants from each continent, are yet fully known. As a result, the following attempt to relate these two sets of data is inevitably tentative. Nevertheless, acceptance of the reality of continental drift has solved several problems that had previously been a puzzle to palaeontologists. For example, the little Lower Permian reptile *Mesosaurus* is found only in two very similar deposits, one in eastern Brazil, the other in South Africa. Though it was apparently an inhabitant of fresh water, it does not appear to have been sufficiently fully adapted to aquatic life to have crossed the 3000 mile wide South Atlantic that today separates these two regions.

Vascular plants first appeared in the Upper Silurian, and woody tree-like plants up to 25 metres high had evolved by the Upper Devonian. Though our information on this early period is still limited, as far as one can tell there were no separate floral realms in the Devonian. Until recently, it was thought that this was also true of the Lower Carboniferous flora; known as the *Lepidodendropsis* flora, this was composed of lycopsids (related to the tiny living club-moss *Lycopodium*), seed ferns and sphenopsids (related to the living horse-tail *Equisetum*). However, it has now become clear that the Lower Carboniferous floras of eastern Asia and of South-East Asia already had some distinctive features.[3]

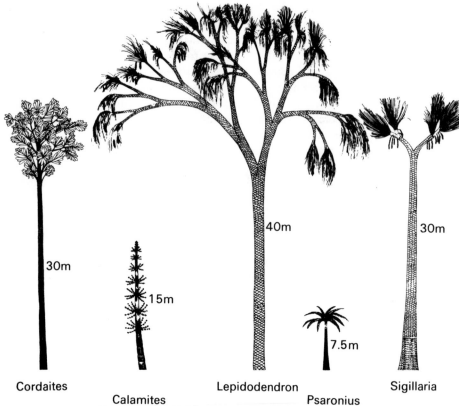

Figure 57. Coal forest trees.

As already mentioned, the great glaciation of all the different parts of Gondwanaland in the Upper Carboniferous suggests that the South Pole lay fairly centrally within that supercontinent at the time. If the re-assembled continents are so positioned, the Equator runs through what are now southern and eastern North America and through southern Europe. It is, then, not very surprising to find that large parts of these areas were covered by swampy, tropical rain forest—an environment rather like that of the Amazon rain forest of today. The absence of dormant buds and of annual growth rings in the fossil remains of this vegetation indicates that it grew in an unvarying, seasonless climate. The flora was dominated by great trees belonging to several quite distinct groups (Fig. 57). *Lepidodendron*, 40 metres tall, and *Sigillaria*, 30 metres tall, were enormous types of lycopod. Equally tall *Cordaites* was a member of the group from which the conifer trees evolved, and the sphenopsid *Calamites* was up to 15 metres high. Tree-ferns such as *Psaronius* grew up to 7½ metres high, and seed-ferns such as *Neuropteris* were among the most common smaller plants living around the trunks of all these great trees.

In the eastern United States and in parts of Britain and Central Europe, the land covered by this swamp forest was gradually sinking.

As it sank, the basins that formed became filled with the accumulated remains of these ancient trees. Compressed by the overlying sediment, dried, and hardened, the plant remains have become the coal deposits of these regions.

Naturally enough, most of the animals known from these coal-swamps were aquatic or semi-aquatic. Little aquatic amphibians such as *Ophiderpeton* and *Microbrachis* are known from both North America and Europe. Larger fish-eating semi-aquatic amphibians, such as *Eogyrinus*, swam about in these waters, around which lived smaller, more terrestrial amphibians, and little lizard-like reptiles such as *Hylonomus*.

Permian and Mesozoic floras   By the early part of the Permian Period, the continents had moved slightly further north, and the areas of Gondwanaland previously occupied by glaciers had become cool-temperate, and had developed the characteristic *Glossopteris* flora described earlier. Now, for the first time, several quite distinct floras can be recognized in different parts of the world.[3] In Europe and North America, which now lay in a drier climatic zone, the rich flora of the moist coal-swamps had disappeared, and lycopod trees (such as *Sigillaria* and *Lepidodendron*) were less common. Instead, the commonest trees were early types of true conifer, whose climatic preference must have been very different from that of their modern, cold-loving descendants. This *Euramerian flora* also included early fern-like plants, seed-ferns, and horsetails. In Asia, two floras can be distinguished, the *Angara* and the *Cathaysia* floras.

Four *floral provinces*, shown in Figure 58, can thus be recognized by the early Permian, but not all the plants were different from one floral province to the next, and many were found in all or several. Also, as might be expected, there was some overlap in the areas between two floral provinces, and these areas may have had a mixed flora. The distinctness of the four major floras lay partly in the fact that some types of plant were abundant in only one flora, such as *Glossopteris* in Gondwanaland, *Cordaites*-like plants and the seed-ferns such as *Angaropteridium* in the Angara flora, and *Gigantopteris* in the Cathaysia flora. The other feature distinguishing the floras was their composition: for example, though conifers were common in the Euramerian flora, they were less common in Angaraland and rare in Cathaysia, the flora of which was dominated by ferns and seed-ferns. It seems likely that these differences were due to the different climates of the regions, that of Gondwanaland being much cooler than that of the neighbouring Euramerian floral province. Similar climatic barriers, caused or reinforced by shallow epicontinental seas and by mountain chains such as the Urals, are the most likely cause of the separation of the northern floral provinces during the Lower Permian.

Between the Lower Permian and the Upper Permian there was a great floral change. Many forms, such as early fern-like plants and the

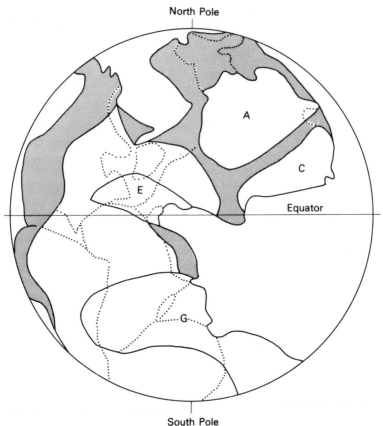

North Pole

Equator

South Pole

Figure 58. The four known floral provinces of the Lower Permian A, Angara; C, Cathaysia; E, Euramerian; G, Gondwanaland. (The floras of the other land areas are unknown.) Dotted lines indicate present-day continental coastlines, shaded areas indicate shallow epicontinental areas.

great trees of the Permo-Carboniferous, disappeared altogether, and their place was taken by new groups such as the osmundas, cycads and ginkgos. This change affected all parts of the world, whose floras therefore became less distinct during the Triassic—though of course the surviving archaic elements in the floras, such as the conifers, retained their former distinctiveness. Florin has reviewed the distribution, from the Upper Carboniferous onwards, of the nine families of conifer and of the taxads (yews), together including over 80 genera.[4] He found that all are restricted to either Laurasia or Gondwanaland, apart from a possible penetration of one Laurasian form into East and South Africa during the Permo-Triassic. This separation was probably mainly climatic, the conifers finding it difficult to spread across the intervening hotter equatorial region, but it must have been reinforced by the intermittent separation of Laurasia and Gondwanaland by shallow epicontinental seas covering southern Europe and northern Africa during the Jurassic and Cretaceous.

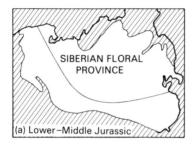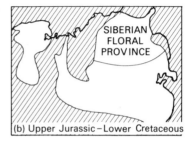

Figure 59. Map of Eurasia, showing modern northern and eastern coastlines and the approximate areas occupied by land and sea, and by the Siberian floral province.

Similarly, within Laurasia, separate floras can be seen which may reflect a persistence of the floral realms of the Palaeozoic, their differences still largely conditioned by climate. Thus most of Northern and Central Eurasia was occupied by a Siberian floral realm[5], which was adapted to cooler climates than the flora of the areas bordering it to the south—Europe, southern Asia and south-east Asia. The boundary between these two floral realms moved northward after the Middle Jurassic, as the sea encroached across western Asia, until in the Lower Cretaceous the more cold-adapted Siberian flora is found only in eastern Siberia (Fig. 59).

Mesozoic animals Though it is possible to identify fairly distinct floras through the Mesozoic, there is little sign of the existence of similar Mesozoic faunas. This may be because such active animals as land vertebrates can quickly traverse a climatic barrier which may provide an insuperable obstacle to the slow spread of plants. It can be no coincidence that the only two really distinctive faunas of today are those of South America and of Australia—both continents which were isolated for a considerable length of time during the Tertiary. For most of the Mesozoic, no such isolated areas existed.[6] In the single Mesozoic land mass, Pangaea, the likelihood of effective separation of two neighbouring regions by barriers such as deserts or mountain ranges is obviously far less than in today's smaller continental blocks. In addition, the Mesozoic mountain ranges in any case lay around the edges of Pangaea, and the climatic differences between the equatorial and the polar regions also seem to have been far less severe than they are today. Though, as already noted, shallow seas did invade the continents, especially during the late Mesozoic, and at times separated Europe from Africa and Asia, these invasions were usually comparatively short-lived and variable in extent, and had little long-term effect on the distributions of animals. As a result of all this there were, until the Cretaceous, no great oceans, mountains, deserts or climatic barriers to divide the Mesozoic land surface into isolated areas within which the faunas could evolve independently in different ways. Instead the amphibians, reptiles and early mammals were able to wander freely over the vast continuous land mass.

The widespread distribution of the Mesozoic faunas can be illustrated by several examples. Perhaps the most striking comes from its beginning, in the Lower Triassic, when a herbivorous reptile called *Lystrosaurus* lived in South Africa, India, Antarctica and western China. In the late Triassic, a peculiar group of herbivorous reptiles known as rhynchosaurs was preserved in rocks in Nova Scotia, Britain, Argentina, Brazil and India. Similarly at that time a tiny early mammal called *Eozostrodon* lived in both Britain and China, and a close relative lived in South Africa. In fact, nearly every family of Triassic amphibian, reptile or mammal found in North America is also known in Europe, and many occurred in South America, Africa and Asia as well.[7] This is also true of the Jurassic and much of the Cretaceous; for example, the largest land vertebrate ever known, the herbivorous dinosaur *Brachiosaurus*, is known from Upper Jurassic rocks in both East Africa and north-western U.S.A., and its close relatives are known also from Europe. Such a widespread distribution was not a special feature of the great herbivores, for fossils of little carnivorous dinosaurs such as *Coelurus* have been found in every continent (except Antarctica), dating from the Lower Jurassic to the Upper Cretaceous.

Neither the dominant groups of animals or plants of the Lower Cretaceous, nor the geography of the world's land areas at that time, were very different from those of the Triassic. Dramatic changes in all three were evident by the Middle Cretaceous, and in the next chapter the transition to the very different world of today will be followed.

References

1 WEGENER A. (1929) *The Origin of Continents and Oceans,* 4th edn. (1966 English translation). Methuen, London.

2 TAKEUCHI H., UYEDA S. & KANAMORI H. (1970) *Debate about the Earth,* 2nd edn. Freeman, Cooper & Co., San Francisco.

3 CHALLONER W. (1973) Carboniferous and Permian floras of the northern continents. In: *Atlas of Palaeobiogeography,* (ed. Hallam), p. 169–186. Elsevier, Amsterdam.

4 FLORIN R. (1963) The distribution of conifer and taxad genera in space and time. *Acta Horti Bergiani* **20,** 121–312.

5 DORF E. (1971) Paleobotanical evidence of Mesozoic and Cenozoic climatic changes. *Proc. N. Am. Paleont. Conv.,* 1969, 323–346.

6 CHARIG A.J. (1971) Faunal provinces on land: evidence based on the distribution of fossil tetrapods, with special reference to the reptiles of the Permian and Mesozoic. In: Middlemiss, Rawson & Newall (eds.) *Faunal Provinces in Space and Time,* special issue of *Geol. J.* **4,** 111–128.

7 Cox C.B. (1973) The distribution of Triassict etrapods. In: *Atlas of Palaeobiogeography,* (ed. Hallam), p. 213–223. Elsevier, Amsterdam.

CHAPTER 7
THE SHAPING OF TODAY

In the seventy million years between the Lower Cretaceous and the Eocene (one of the Epochs of the Tertiary Period, see Figure 49) the world changed from a single land-mass dominated by gymnosperms and reptiles to a pattern of separating continents dominated by flowering plants and mammals. Because the rise of the flowering plants took place in the early Cretaceous, before that of the mammals in the Paleocene, the effects of continental drift upon the distributions of these two groups were rather different.

Continental drift facilitated the development of separate, distinctive faunas and floras, not merely because of the physical separation of the new continents by ocean barriers, but in other ways also. The climates of land areas newly bordered by seas became milder and less variable. Where new mountain ranges lay across the path of the prevailing rain-bringing winds, new deserts grew in their lee. Finally, as the continents continued northward, their northern fringes reached such a high latitude that they became covered by permanent ice-sheets. This may have been the reason for the exaggeration and narrowing of the climatic zones; it may also have led in turn to the great Ice Ages of the Pleistocene, which wrought havoc upon the plant and animal life of the Northern Hemisphere. It is, perhaps, no mere coincidence that both the Permo-Carboniferous glaciation and the Pleistocene glaciations occurred at times when a considerable area of land lay near to one of the poles.

It is worth considering for a moment what patterns of distribution we might expect to find had the continents always had their present positions, so that the only changes would then have been the relatively minor climatic variations of the Northern Hemisphere Ice Ages, and changes in sea level making or breaking the intercontinental Bering and Panama land bridges. At times when the climate was warmer than it is today, the spread of animals and plants across the Bering region between Siberia and Alaska would have been possible. Similarly, in the absence of the deserts of the Middle East, there would also have been a single tropical fauna and flora stretching from West Africa to South-East Asia. It would not be surprising, however, if the later development of these deserts, dividing the tropical region into African and Asian sections, had allowed distinctive features to appear in the faunas and floras of each. Finally, it might have been expected that the complete isolation of Australia and the almost complete isolation of South America would have led to the development of unique faunas and floras on these continents.

The rise of the flowering plants Though these features are clearly visible in the accepted patterns of animal and plant distribution, other aspects of these patterns are less simple to explain. Today's floral realms are based upon the distribution of the flowering plants, or angiosperms, which have been described by Good[1] (Fig. 60). Perhaps the most fundamental feature of angiosperm distribution is the fact that, almost everywhere in the world, four families are among the six most numerous

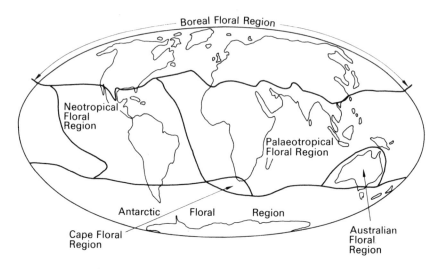

Figure 60. The floral regions of the world today. (After Good, 1964; courtesy of Longmans, Green and Co. Ltd., London.)

—the Compositae, Graminae, Leguminosae and Cyperaceae. Similarly, dicotyledonous angiosperms are almost everywhere more abundant and diverse than the monocotyledonous types. Even the floras of the isolated continents, Australia and South America, though they certainly display unusual features, are not basically unique, composed of major groups found nowhere else.

Furthermore, if the spread of angiosperms through the world had been through the pattern of continents we see today, one would expect each of the three southern temperate regions (Australia and the southern parts of South America and Africa) to have a flora derived from that of its own adjacent tropical region. This is not the case, as is shown most clearly by the contrast between the floras of Australia and of New Guinea. Twelve of the 28 most dominant families of angiosperm in each region are completely absent from the other region, and the Australian flora appears as an intrusive element in an otherwise uniform southern Pacific flora. In fact, rather than being related to the floras neighbouring them to the north, the floras of the three south temperate regions show considerable similarity to one another, despite their separation by wide expanses of ocean. For example, over 700 species of angiosperm are more or less entirely restricted to two or three of these regions, and six families (the Cunoniaceae, Escalloniaceae, Gunneraceae, Philesiaceae, Proteaceae and Restionaceae) are found in all three. Even more well defined is the flora of the extreme cold temperate region south of 45°s, which is composed largely of groups of angiosperm which are scarcely, if at all, represented further north. This "Antarctic" flora is characteristic of the extreme southern end of South America, southeast Australia, Tasmania and New Zealand—it is absent from South Africa, which extends only to 35°s. This type of distribution has long puzzled biogeographers—see, for example, Darlington's discussion of the distribution of the southern beech, *Nothofagus*.[2]

All these facts clearly suggest that the angiosperms spread through the world at a time when the continents had not yet split apart, so that a flora of fairly uniform composition (at family level) spread everywhere. Part of this uniform flora became adapted to the temperate climate of the southern part of Gondwanaland. When this supercontinent split up, the flora found itself on three separate continents and, though some genera have since become extinct, many are still to be found in more than one of these areas. This theory is compatible with what is known of the sequence of events.

The angiosperms originated in the Jurassic, were still a relatively insignificant part of the floras of the Lower Cretaceous, and became dominant in the Mid-Cretaceous.[3] Gondwanaland is thought to have broken up during the Cretaceous and, though the precise timing is still uncertain, it is clearly possible that the angiosperms spread throughout that supercontinent before the break-up—some types still characteristic of the south temperate region today, such as the Proteaceae and the southern beech, *Nothofagus*, had already appeared in Australia in the Mid-Cretaceous. In Laurasia, two separate angiosperm floral realms had already appeared by the Upper Cretaceous.[4] One covered eastern North America and the whole of Europe, the other extended throughout Asia and western North America. The two were separated by the shallow epicontinental Turgai Straits, which lay east of the Urals and ran across Asia from north to south, and this region still marks the separation between the European and the Asiatic sub-regions of the Boreal floral region today.

Though it is not yet possible to trace in detail the transition between the two, the basic characteristics of today's floral realms are already discernible in the Cretaceous floral distributions.

The mammals supplant the reptiles Though the earliest mammals appeared in the Triassic, long before the earliest angiosperms, it was not until the late Cretaceous that they became more varied, while their differentiation into the many orders seen today took place only after the extinction of the dinosaurs at the very end of the Cretaceous. Thus, though in the Upper Cretaceous there was only one order and family of marsupials, with nine known genera, and four orders of placental mammal, with five families and eight known genera, by the Upper

Paleocene, only ten million years later, there were now two families and eleven genera of marsupials and 18 orders, 44 families and over 80 genera of placentals.

Because the evolutionary radiation of the mammals did not take place until after the break-up of Laurasia and Gondwanaland had commenced, and because today's zoogeographical regions are based upon the distribution of the mammals, this pattern is more clearly determined by modern geography than is that of the angiosperms (Fig. 61). (This may explain why those scientists who worked on mammalian palaeontology were for so long firmly opposed to the theory of continental drift.) The pattern of vertebrate distribution has been fully documented and discussed by Darlington.[5]

The influence of the continental break-up is seen at the largest scale in its effect on the distribution of the marsupials. These more primitive mammals (whose young leave the uterus at a very early stage and then complete their development in the mother's pouch) were apparently the first to appear—probably some time in the Lower Cretaceous. Later, the more advanced mammals, the placentals (in which the whole period of embryonic development takes place in the uterus), underwent a rapid period of evolution into many types, which spread over most of the world and virtually replaced the marsupials.

Had the placentals spread to every corner of the world before the break-up of the supercontinents, then the marsupials would almost certainly have been entirely wiped out. Instead, the marsupials survived on different continents for different lengths of time, and so it must have been after the supercontinents had started to break up that the great evolutionary expansion of the placentals occurred.

It is still difficult to be sure of precisely what did take place, because our knowledge of the history of mammals in the Cretaceous and early Tertiary (especially in Australia and Africa) is still very incomplete. The following theory, however, is probably accurate in most respects.

Australia Both marsupials and placentals are known to have existed in North America in the Upper Cretaceous. Marsupials, but *not* placentals, colonized Australia, which lay at the far end of the Mesozoic supercontinental assembly. This suggests that the marsupials, but not the placentals, had evolved and spread to Antarctica–Australia before this mass became isolated from the rest during the early Cretaceous.[6] Later, Australia and Antarctica split apart, and Antarctica moved to the extreme southern position that has rendered it virtually devoid of life. Australia remained unconnected with the other land-masses, but rats and bats (both placental groups) eventually reached it along the stepping-stones of islands from south-east Asia. Man has in the last few thousand years introduced the dog and, even more recently, several other placentals, including the rabbit. Apart from these, Australia has no placental mammals. As a result, it still retains a marsupial fauna which has been able, over many millions of years, to radiate into a great

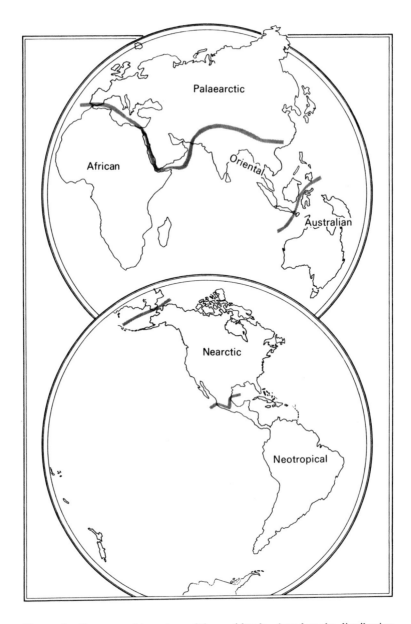

Figure 61. Zoogeographic regions of the world today, based on the distribution of mammals.

variety of forms occupying the niches that placentals have filled everywhere else. Marsupial equivalents of rats, mice, squirrels, jerboas, moles, badgers, ant-eaters, rabbits, cats, wolves, and bears all exist, and look superficially very like their placental counterparts. Only the place of the herbivorous placental ungulates (the hoofed mammals—such as rhinoceroses, horses, and deer) has been taken by marsupials that look quite unlike them, for this niche is occupied by the kangaroos and wallabies.

133

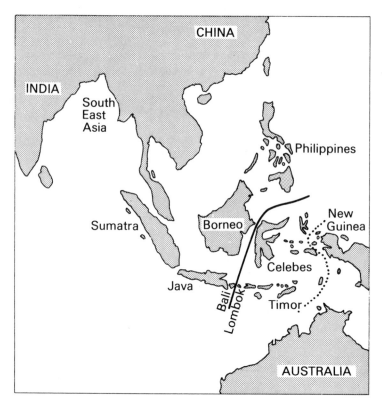

Figure 62. The continuous line marks Wallace's Line, which was along the edge of the continental shelf of South-East Asia. The dotted line marks the edge of the Australian Continental shelf.

Since Australia reached its present position, its marsupial fauna has spread westward along the chain of islands toward south-east Asia. Some placental mammals of the Oriental region have similarly spread eastward along these islands, until there is a little overlap of the two faunas in the region between Java and New Guinea. The line of separation—often known as *Wallace's line*—between the Australian zoogeographic region and the Oriental region (India and south-east Asia) is not, therefore, sharp and precise, but it is usually drawn along the edge of the continental shelf of south-east Asia (Fig. 62).

South America   Though marsupials are known from the Upper Cretaceous of South America, the first record of its placental fauna is in the Upper Paleocene. It then included six different endemic orders belonging to two groups, the primitive ungulates and the edentates.[7] These were joined in the Lower Oligocene by primates and rodents. It is significant that the early Tertiary fauna of Africa, too, contained distinctive primitive ungulates and was colonized by primates and rodents in the Lower Oligocene. These facts suggest that a few mammals were occasionally able to reach South America from Africa. However, the connection clearly did not permit frequent passage, since the faunas of the two continents otherwise evolved along independent lines. It therefore seems likely that the migration route was not a land bridge but was a chain of islands, rather like the East Indies of today. Whatever its form, this pathway to South America finally became impassable after the Lower Oligocene.

Due to this almost complete isolation during the early Tertiary, many unique animals evolved in South America. Of these, only the armadillos, the tree-sloths, and the ant-eaters survive today, together with the opossum as the lone relic of the marsupials. The New World monkeys (such as the spider-monkeys and the howlers) and the entire group of rodents known as *caviomorphs* (for example, the capybara, the guinea-pig, and New World porcupine) are the only survivors of South America's unique primitive placentals. The rest, including all the primitive South American ungulates (such as *Thoatherium* and *Toxodon*), were unable to survive the competition from the more advanced placentals that entered South America when, toward the end of the Pliocene, it at last became permanently connected with the rest of the world via the new Panama Isthmus to North America. Most of this narrow bridge between the continents consists of tropical lowlands, similar to those of South America; as a result, most of the isthmus has been colonized mainly by South American animals. The division between the faunal regions of South America (the Neotropical region) and North America (the Nearctic region) lies just south of the cool Mexican Plateau, which these South American tropical animals have been unable to colonize.

Except in Australia and South America, competition from the more advanced placentals that evolved later in the Tertiary led gradually to the complete extinction of the marsupials; in North America and Eurasia they did not survive beyond the Miocene. The continued splitting of the continents during the Tertiary was therefore not the main factor in differentiating the faunas of North America, Eurasia, Africa, and India. Instead, as we shall see, it was the gradually cooling climate of the northern continents (due partly to their movement toward the North Pole) that caused the great difference between the faunas of these continents and those of the Ethiopian region (Africa and Madagascar) and the Oriental region.

Old World tropics—the Oriental and African regions The Oriental region includes both South-East Asia, which may always have been part of Laurasia, and India—originally part of Gondwanaland. If India became separate at about the Cretaceous/Tertiary boundary it must originally have had a fauna of early mammals and, perhaps, dinosaurs. Unfortunately the Lower Tertiary vertebrate fauna of India is still unknown and it is not until after it became connected with Asia in the Miocene and was invaded by the Asian tropical fauna that the vertebrate fauna of India is known.

The early Tertiary faunas of Africa, on the other hand, are better known,[8] though even here the record only begins with the Upper Eocene to Lower Oligocene fauna of the Fayum, Egypt. This fauna contains two different groups of placental. One group consists of four endemic orders of primitive ungulate—elephants, hyracoids (conies), sirenians (sea cows) and the extinct embrithopods. These are the descendants of an early ungulate stock which managed to enter Africa from Europe in the Paleocene or Lower Eocene, probably across a shallow Mediterranean-like sea which kept out the other early placentals. The other group of Fayum placentals had apparently entered Africa only recently; this includes artiodactyls, creodonts, insectivores, rodents, and also the earliest members of the anthropoid (apes and man) primate line, whose later evolution was centred in Africa. Though some interchange with Eurasia took place in the Miocene, there was still no wholesale exchange between the two continents, and new endemic African forms evolved, such as Cape golden moles, tenrecs and elephant shrews.

Because the African and Oriental regions are the only two Old World areas in which tropical mammals can exist today, it is not surprising to find a number of similarities between their faunas. Primitive primates such as the lemurs and lorises, as well as Old World monkeys and porcupines, apes, rhinoceroses, elephants, and the pangolin (*Manis*), are all found exclusively in these two areas. But in nearly every case, each group is represented by slightly different forms in the two areas. For example, the African rhino, elephant, and porcupine all belong to genera different from those found in the Oriental region. Similarly, the lemurs of Madagascar and the chimpanzees and gorilla of Africa are not found in the Oriental region, where these groups are represented by the lorises and by the orang-utan and gibbon. The two regions are, perhaps, sufficiently far apart for these differences to be expected, even if animals could have roamed freely through tropical forests all the way from western Africa to India. In fact, the appearance of deserts in North Africa and the Middle East has split the Old World tropical fauna into two sections, which will probably become increasingly different from one another. This aridity, which began to appear in mid-Tertiary times, has had a further effect upon the fauna and flora of Africa, for in East and Central Africa it has led to the replacement of the original forests by drier grasslands called *savannah*. The huge herds of ungulates that graze on these grasslands, such as the many types of buck (impala, gazelle, gnu, antelope, and others), giraffes, buffalo, zebra, and wart-hogs are now thought of as the "typical" fauna of Africa; but in reality these are late-comers to the African scene. The appearance of this savannah has helped even further to isolate the original forest-fauna of Africa from its Indian relatives. This forest fauna is now restricted to western and west central Africa, and to Madagascar, which was linked to Africa in the early Tertiary, though it is now separated from the mainland by 240 miles of sea. Perhaps while still joined to Africa, Madagascar received a forest-fauna of primitive members of several groups (primates, insectivores, rodents, and carnivores) that still survive there in isolation, although some species are now very rare.

Holarctic region—North America and Eurasia The faunas of Eurasia (the Palaearctic region) and North America (the Nearctic region) are in many ways very similar to one another, and are sometimes considered as a single "Holarctic" region. Their similarities are due partly to their climate—they include nearly all the temperate and cold regions of the animal and plant world—and partly to the fact that, for much of their history, it has been possible for animals to migrate from one to the other. North America, Europe and Asia were connected in two different patterns in the Tertiary.

Until the end of the Lower Eocene, the North Atlantic had not yet joined the Arctic Ocean, and North America was still directly connected to Europe via Greenland and Scandinavia. North America provides our best record of Cretaceous and Paleocene mammal faunas, and appears to have been the centre of early placental evolution. Though the European Tertiary fauna is not known until the Upper Paleocene, it is then very similar to that of North America; nearly all the European families are also known in North America, and some genera are found in both continents. On the other hand, a number of North American Upper Paleocene families are unknown in Europe, and the climate of the northern connection between the two continents appears to have acted as a filter.

Asia during this time was isolated from Europe by a seaway known as the Turgai Straits, lying east of the Urals. The Bering connection between Siberia and Alaska therefore provided the only route for exchange between Asia and conjoined North-America/Europe. However, the Bering region appears to have lain at a fairly high latitude, and its cold climate is probably the reason why many mammalian groups known in North America, such as the marsupials, failed to cross to Asia.

The second pattern of relationship between the northern continents began in the Mid Eocene. The North Atlantic now separated Europe from North America, and such new European groups as palaeothere horses and primitive relatives of the camel could not cross to North America. On the other hand, the drying-up of the Turgai Straits now allowed the European fauna direct entry into Asia. At the same time, the climate appears to have become warmer, so that a greater variety of mammals was able to enter Asia from North America via the Bering connection.

This Bering area was now the only link between Eurasia and North America, and from henceforth it was climate, not continental movements, that determined the faunal relationships between these two continents. When the climate became cool, as in much of the Oligocene, few mammals crossed. When it improved again in the Lower Miocene, a number of Asian mammals migrated to North America. The final deterioration began in Mid Miocene times. From then on, most migrants are large forms and, even more significantly, types which are tolerant of cooler temperatures—such warmth-loving forms as apes and giraffes could not reach North America. This climatically based exclusion became progressively more restrictive, until in the Pleistocene only such hardy forms as the mammoth, bison, mountain sheep, mountain goat, musk-ox and man himself were able to cross. The final break between Siberia and Alaska took place 13,000 to 14,000 years ago.

Despite the long history of intermittent connection between the Nearctic and Palearctic regions, each has certain groups of animals that have never existed in the other one, and also groups that did reach both regions, but became extinct in one and were never replaced by subsequent colonization from the other. Pronghorn antelopes, pocket gophers and pocket mice, and sewellels (the last three groups are all rodents) are unknown in the Palearctic region, whereas hedgehogs, wild pigs, and murid rodents (typical mice and rats) are absent from the Nearctic region. The domestic pig has been introduced to North America, as have mice and rats at various times. The horse became extinct in the Americas during the Pleistocene Period, but had crossed the Bering connection to Eurasia. Horses were therefore unknown to the American Indians until they were introduced by the Spanish *conquistadors* in the 16th century.

Both continents were stripped of virtually all tropical and subtropical animals and plants by the Pleistocene Ice Ages. This happened so recently that the faunas and floras as yet have had no time to develop any new, characteristic groups. Since they also have no old relict groups, such as the marsupials, it is the poverty and the hardiness of their faunas that distinguishes them from those of other regions. Many groups of animals are absent altogether and, of the groups that are present, only the more hardy members have been able to survive. Even these become progressively fewer toward the colder Arctic latitudes. In North America there is, in addition, a similar thinning out of the fauna in the higher, colder zones of the Rocky Mountains. This is a general feature of the fauna and flora of high mountains, as described in Chapter 2.

Almost completely isolated from the warmer lands to the south by the Himalayas and by the deserts of North Africa and southern Asia, the Palearctic fauna has received hardly any infiltrators to add variety. The Nearctic region, on the other hand, is linked by the Panama Isthmus to South America, from which some animals have been able to spread. For example, the opossum, armadillo, and New World porcupine have colonized North America, as have a number of birds, such as the humming birds, mocking birds, and New World vultures.

Fossil floras and changing climate   The catastrophic changes that overtook North America and Eurasia during the Ice Ages were only the culmination of a progressive cooling of world climates that began in the Cretaceous and continued throughout the Tertiary. Evidence of this is provided by a number of fossil floras from the Northern Hemisphere (far less is known of the Tertiary floras of the Southern Hemisphere).[9] For example, a series of floras from about 70°N in Alaska, ranging through about 30 million years of the Middle Cretaceous, shows that both the climate and the flora changed considerably. The earliest of these floras contains the remains of a forest dominated by ferns and by gymnosperms such as cycads, ginkgos, and conifers. The nearest living relatives of this flora are found in forests at moderate heights in warm-temperate areas at about 25–30°N—for example, in south-east Asia. By the time of the last of the Middle Cretaceous Alaskan floras, the flora had changed in two ways. First, the angiosperms had by this time diversified to such an extent that they dominated the flora. Second, this flora contains the remains of forest similar to that found today at a latitude of 35–45°N in the region of North China and Korea—much further north than the living relatives of the earlier flora. The differences between these successive fossil floras therefore suggest that the climate of northern Alaska was already becoming cooler in the Middle Cretaceous.

Dramatic climatic changes have occurred also in Europe, where a slightly later (Upper Cretaceous) flora contains tropical mangroves. During the Lower Eocene, about 50 million years ago, tropical conditions still extended at least as far north as southern England. This is clearly shown by the discoveries of fossil seeds and fruits of about 100 genera of plants preserved in the London Clay (of Lower Eocene age) that makes up the low cliffs and the foreshore of the Isle of Sheppey, near the mouth of the River Thames. Of these genera, 73 have living relatives, nearly all of which are found in the tropics, many of them in Malaysia and Indonesia. The flora includes plants such as the palm-trees *Nipa* and *Sabal*, cinnamon, magnolia and *Sequoia*. The accompanying fauna of the London Clay, which included crocodiles and turtles, also indicates a tropical climate.[9]

In both Europe and North America, the transition to the Oligocene was accompanied by further cooling of the climate, so that specifically tropical types of plant disappeared and were replaced by forms more tolerant of cool climates, as well as by forms found only in cool climates.[9,10] This trend continued throughout the Miocene, but a more abrupt change seems to have taken place in the Pliocene. This change is perhaps most clearly shown by the fact that, though only 10 per cent of the species in the Lower Pliocene flora of Europe still occur there today, the figure for the Upper Pliocene floras is over 60 per cent.[10]

By the end of the Pliocene, then, the flora of the northern continents was essentially a modern one. But the period of time between then and today has seen not a mere continuation of the cooling tendency shown by the rest of the Tertiary, but a series of dramatic cyclical changes.

Pleistocene ice ages   During the Pleistocene Period, the temperate climate of northern Europe was periodically transformed into an Arctic one, and glaciers began to form in highland regions, gradually spreading out over the lower-lying areas.[11] Temperate forests retreated before them and were replaced by boreal forest, then by tundra; finally the land was covered by a lifeless sheet of ice.

At various stages during the Pleistocene, ice covered Canada and parts of the United States, northern Europe, and Asia, and most of the British Isles apart from the region south of the Thames. A number of geological features that we can see today provide evidence of such past glaciation—one of the most conspicuous of these is the *glacial drift deposit, boulder clay*, or *till* covering large areas and sometimes extending to great depths. This is usually a clay material containing quantities of rounded and scarred boulders and pebbles, and geologists consider it to be the detritus deposited during the melting and retreat of a glacier. The most important feature of this till, and the one by which it may be distinguished from other geological deposits, is that its constituents are completely mixed—the finest clay and small pebbles are found together with large boulders. Often the rocks found in such deposits originated many hundreds of miles away, and were carried there by the slow-moving glaciers. Fossils are rare, but occasional sandy pockets have been found that contain mollusc shells of an Arctic type.

Many of the valleys of a hilly glaciated area have a distinctive smoothly rounded profile, because they were scoured into that shape by the abrasive pressure of the moving ice. Other evidence for former glaciation is less dramatic. When the land was covered by ice sheets, the water in the soil froze and expanded greatly, raising the surface of the ground into domes and ridges: this process is known as *frost-heaving*. Stones that lay on top of these ridges slid down the ice into the hollow between them, and so became arranged in the *stone stripes* and *stone polygons* that have been detected by aerial photography in certain districts of Britain, such as East Anglia (Fig. 63).

Areas south of the glaciated regions experienced what has been termed a *periglacial* climate, which involved Arctic conditions and soils constantly disturbed by frost action. These areas probably had scanty, treeless, Arctic vegetation. Still further south, many of the areas of North America and North Africa today covered by desert once had high rainfall, and freshwater lakes occurred there. Evidence for the existence of these lakes in past times is provided not only by the geological deposits, but also by the present-day distribution of certain freshwater animals. In western Nevada there are many large lake basins that are now nearly dry, but in the remaining water-holes there live species of fish that belong to a fauna once widespread in the area.[12] Had these *relict* populations (page 14) been separated for a very long period, they would have begun to diverge in their genetic make-up from the main population. But because they are still very similar, the two populations must have been interconnected by an extensive body of water in the geologically recent past. The pluvial periods of the Quaternary

Figure 63. Polygons in Arctic Canada.

provided the conditions necessary for these aquatic animals. An Old World example is provided by the waterbug genus *Corixa*, which is now widespread in Europe, but in Africa exists in only a few scattered localities extending south to the Rift Valley in Kenya. Presumably it was much more abundant when what is now the Sahara was dotted with lakes, and has since become restricted in distribution as the climate has grown drier.

Glacials and interglacials The Quaternary Era, however, has not been one long cold spell. Geologists have for many years been aware that the glacial drift that covers much of northern Europe, for example, is not one homogeneous deposit. In many regions it can be seen to be divided into layers, varying greatly in thickness. Much work has been carried out on the deposits of East Anglia, and here certain layers have been described bearing fossil remains that do not belong to an Arctic climate. Perhaps the best-known of these is the *Cromer Forest Bed*, a dark, peaty band, often 1·5 metres thick, exposed beneath glacial drift on parts of the Norfolk coastline. Freshwater sediments containing fossils of Arctic flora and fauna lie above and below this layer.

The band itself, however, contains pollen from such deciduous trees as oak, alder, lime, elm, and hazel, which are characteristic of north temperate climates and cannot survive under colder conditions. In the lower and upper parts of the forest bed most of the pollen is of pine, spruce, and birch trees, which live in colder *boreal* (northern) climates. This suggests that Arctic conditions gave way to a warmer climate for

141

some time before becoming colder again. During the period over which the Cromer Forest Bed[13] was formed, then, the climatic fluctuation probably took the form of Arctic—Sub-Arctic—Boreal—Temperate—Boreal—Sub-Arctic—Arctic. Each climatic stage in this fluctuation lasted long enough to allow the vegetation best suited to it to develop fully. Thus a treeless Arctic tundra vegetation was replaced by a sub-Arctic flora with dwarf birch and dwarf willow, followed by boreal forest, mostly of conifers such as pine and spruce, together with birch; then came temperate deciduous forest much like that of present-day northern Europe with trees such as oak, elm, and hazel, this being replaced by boreal forest, then sub-Arctic scrub, and finally Arctic vegetation as the climatic cycle ended. These differences in the types of vegetation found in East Anglia during the Lower Pleistocene, which tell us what the climate was like at the time, are known today from identification of the pollen from the different plants involved found at different layers in the Cromer Forest Bed and similar deposits.

Such a series of climatic changes within the Quaternary Era is known as an *interglacial*. The Cromer deposit is just one of a number showing climatic, and hence vegetational, changes as described above. Geologists have been able to distinguish four major periods of glaciation within the Quaternary, separated by three major interglacials.

Different names have been given to the glacial and interglacial periods in the different parts of the world where they have been distinguished. Figure 64 shows the chronology of glaciation obtained from European regions and from North America. R. G. West, whose system of classification for Britain is given in the table, has suggested that there were at least three cold periods with intervening milder periods prior to the Cromerian, although no glacial deposits from these periods have been found in Britain—the evidence for these pre-Cromerian climatic changes comes from pollen analysis. The names for the stages in West's classification are taken from the places in East Anglia from which they were first definitely described. It is tempting to speculate that the major glacial stages recorded from the other parts of the world coincide with the four described by West. Although this may be the case, it is not yet proven and is perhaps best not implied; hence the local names are retained.

Interstadials—the Allerod period   The climatic oscillations involved in this glacial/interglacial cycle were not smooth, regular ones. The glacial periods themselves were interrupted by milder stages, termed *interstadials*. These are not the same as interglacials; they are periods when the climate became warm enough to support boreal forest, but was too cold for temperate forest. The final (Weichselian) glaciation was probably interrupted by at least six interstadials, two of which occurred at the end of the period. Of these two, the most widespread is the Allerod interstadial, named after a town on the Danish island of Sjaelland, to the north of Copenhagen, where it was first identified.

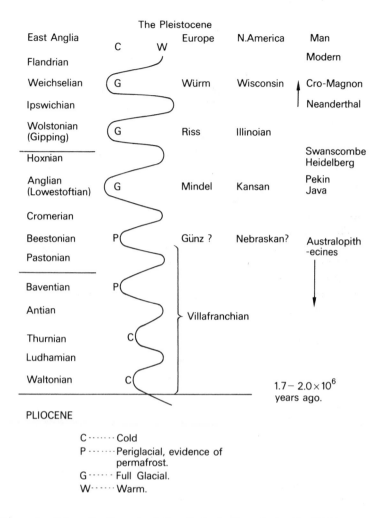

The Pleistocene

East Anglia			Europe	N.America	Man
	C	W			Modern
Flandrian					
Weichselian	G		Würm	Wisconsin	Cro-Magnon
Ipswichian					Neanderthal
Wolstonian (Gipping)	G		Riss	Illinoian	
Hoxnian					Swanscombe Heidelberg
Anglian (Lowestoftian)	G		Mindel	Kansan	Pekin Java
Cromerian					
Beestonian	P		Günz ?	Nebraskan?	Australopith -ecines
Pastonian					
Baventian	P				
Antian			Villafranchian		
Thurnian	C				
Ludhamian					
Waltonian	C				

$1.7 - 2.0 \times 10^6$
years ago.

PLIOCENE

C ········ Cold
P ········ Periglacial, evidence of permafrost.
G ········ Full Glacial.
W ········ Warm.

Figure 64. Schematic diagram of climatic fluctuations during the Pleistocene period in Britain (based on the data of R. G. West) and possible correlations with European and North American climatic fluctuations.

In many parts of northern Europe, the retreating glaciers left extensive lakes behind them. Valleys were often blocked by moraines (mounds or ridges of till left by the melting ice) and lakes formed behind these natural dams. These lakes were cold and clear and contained little life. During the warm Allerod interstadial, however, there was a great invasion of plant and animal life as a result of the rapidly rising temperatures. This sudden explosion of life was checked by a subsequent return to cold conditions before the final overall increase in temperature of the post-glacial period. However, the lake sediments formed during this period of changing climate contain much evidence of these biotic changes. The layers deposited during Allerod times are often rich in organic matter and frequently contain fossil fruits, seeds, and pollen of the more warmth-demanding species then found in northern Europe. Evidence from plant fossils suggests that during this interstadial the vegetation developed no further than scrubland with some continuous woodland of birch (*Betula pubescens* and *B. pendula*), and poplar (*Populus tremula*), which need somewhat warmer conditions than does the dwarf birch (*B. nana*). However, these plant remains do not provide very precise information about the rapid temperature changes at this time. Insects—which, unlike plants, are mobile—seem to have been able to respond more quickly to a change in temperature than could the vegetation, and may therefore provide a more precise indication. For example, the presence in British Allerod deposits of fossils of some beetle species now typical of Mediterranean regions suggests that Britain may then have been warmer than it is today.[14]

The layers of clay below (deposited before the Allerod) and above (deposited after the Allerod) are largely of a mineral origin, and in these such fossils are scarce and are mainly of plants now associated with cold conditions. Some of these plants are no longer native to the British Isles and are restricted today to Arctic or alpine regions; they were probably unable to tolerate the increasingly warm climate after the Weichsel glaciation, although competition from new plant species that then invaded Britain may have contributed to their disappearance. Other cold-loving plants, such as the mountain avens (*Dryas octopetala*) and the dwarf birch (*Betula nana*), still remain in Britain as relicts, mainly in high mountain areas. During the cold periods immediately before and after the Allerod interstadial, both *Dryas* and the dwarf birch were widespread in the lowlands of Britain and the rest of north-west Europe. Today *Dryas* seems to be restricted to those areas in which the summer temperature does not exceed 27°C—this gives an indication of the summer temperatures in north-west Europe during the cold periods of late Weichselian times. During these cold periods, the vegetation was probably of grass-sedge tundra with dwarf trees or shrubs (dwarf birch, arctic willow, and juniper), and the Arctic-alpine plants described above.[15]

Life after the Allerod   The cold period following the Allerod lasted only about 500 years, up to about 8300 B.C. Evidence shows that after this the climate became once again warmer, and this time for a longer period. In the layers of fossilized lake deposits (the *stratigraphic profile* of the lake), mineral deposits that had slumped into the ancient lakes as a result of frost movements and soil erosion are succeeded by organic muds that indicate an abundant growth of warmth-demanding aquatic plants. Often the rapid deposition of the remains of these plants led to lakes becoming filled with organic detritus and choked by invading sedges and reeds, frequently followed by trees, particularly birch (*Betula*). Later the warm climate grew wetter and these swampy birch woods were destroyed by the rapid and vigorous growth of bog-mosses (*Sphagnum*).

The increasing warmth since the close of the last glaciation has had a considerable effect on the vegetation of the Northern Hemisphere. The belt of temperate deciduous woodland that was forced south during the glacial periods spread northward again, following the retreat of the glaciers. Certain trees, such as wingnut (*Pterocarya*), hemlock (*Tsuga*), and the tulip tree (*Liriodendron*), which were present in the European deciduous forests of pre-Quaternary times, disappeared after the Weichselian glaciation. Presumably, these trees disappeared in Europe as the deciduous forest belt had its southward retreat blocked by the Alps and the Pyrenees. By contrast, in North America the main mountain ranges run roughly north-south and hence did not interfere with the southward spread of these trees during glacial times. These could therefore return northwards when the climate later became warmer and, as a result, all three genera are still native to the hardwood forests of North America. Thus, in terms of diversity of species, it was a much poorer forest that reinvaded the tundra regions of northern Europe following the retreat of the glaciers. Other factors tended to reduce the richness of the floras of certain areas still further. The melting ice caps caused a rise in sea level, which eventually led to the submergence of the wide tract of land that up to this time had linked Britain to the rest of Europe. This occurred about 5500 B.C., and many plants with slow migration rates had still not crossed into Britain and were thus permanently excluded from the British flora. The separation of Ireland from the rest of Britain also occurred during this period, and many species native to Britain had not established themselves as far west as Ireland. As a result, plants such as the lime tree (*Tilia*) and Herb Paris (*Paris quadrifolia*) are not found growing wild in Ireland.

There is, however, one group of plants of great interest to plant geographers that did succeed in reaching Ireland before the rising sea level separated that country from the rest of the British Isles, and this is known as the *Lusitanian* flora. Lusitania was the name for a province of the Roman Empire consisting of Portugal and part of Spain and, as its name suggests, this flora has affinities with that of the Iberian peninsula. Some of the plants—such as the strawberry tree (*Arbutus unedo*) and giant butterwort (*Pinguicula grandiflora*)—are not found growing wild in mainland Britain; others, such as the Cornish heath (*Erica vagans*) and the pale butterwort (*P. lusitanica*), are found in south-western England as well as in Ireland. It therefore seems likely that these plants spread from Spain and Portugal up the Atlantic seaboard of Europe in post-glacial times but were subsequently cut off by the rising sea levels (see pages 16–17).

Post-glacial climatic changes   Much information regarding post-glacial changes in Europe has been obtained by the analysis of pollen from lake and peat deposits.[16,17] This provides a continuous record of vegetational changes over many millennia, but has certain disadvantages. First, pollen may travel considerable distances and hence have its origin far from the site of fossilization. Second, plants producing large quantities of air-borne pollen will be over-represented in all samples and hence may give an impression of abundance. Finally, pollen identification can often be taken only to genus level, and sometimes only to family level. It is however much easier to identify *macroscopic* plant fossils (those visible to the naked eye) such as fruits and seeds, but these are mainly restricted to those of marsh and aquatic plants that grew locally.

Further problems arise when one attempts to use these fossils as indicators of specific climatic conditions. One could assume that the fossil type had precisely the same environmental requirements as the living type, but even if this were so, there are very few instances where the precise environmental requirements of a living species are known. Thus any reconstruction of Quaternary climates based solely upon the fossil evidence must be a very tenuous one. It is usually possible to come to only vague conclusions—for instance, that temperate forest types indicate a temperate climate—but even this may be a dangerous generalization, a case in point being early estimates of Allerod temperatures, which were too low.

Recently, a great deal of research has been carried out on deep sea sediments containing fossil Foraminifera (Fig. 65). These are minute oceanic animals, some of which are free-floating (planktonic), while others live on the sea floor. They mostly secrete calcareous shells, which in the planktonic forms fall to the ocean bed when they die and, together with the skeletons of the bottom-dwelling types, form extensive deposits. Many of these "Forams" can be identified with great accuracy, and so fossil forms can often be dated. They can also be placed in time by using radio-carbon dating (a method used with great success to date many fossils), which depends on the constant rate of decay of radio-active carbon-14 into "ordinary" carbon-12; this rate is known and therefore the fossil's age can be determined from the amount of the radio-active isotope remaining in it. Also, since certain Forams have quite precise water temperature requirements, we have a good idea of the temperature at the time when they were deposited.

By examining long cores of marine deposits (collected by lowering weighted tubes, fitted with suction pistons, into the soft sediment) geologists were able to work out a continuous record of the climatic changes throughout the Pleistocene Period, and this was then correlated with the results from terrestrial deposits. By analysing the ratios of oxygen isotopes in deep-sea sediments, long-term changes in ocean temperature have been detected. The ratio of the isotopes Oxygen-18 (heavy oxygen) and "ordinary" oxygen-16 in the carbonates of biological origin in marine sediments (such as those from the fossil Forams) shows a close relationship to the temperature of the water in which the carbonate was formed.[18] This method of detecting past climates does not present the same problems as those encountered when working with fossils on the land. Certain other difficulties do arise, however. For instance, ocean temperatures are far less subject to

Figure 65. *Foraminifera.* Microscopic organisms of largely marine and estuarine origin which have been useful as indicators of past ocean temperatures and salinities.

fluctuation as a result of climatic change than are those on the land. Evidence of such variations is thus scaled down, especially that of a short-term change such as the Allerod. Also we have to assume that marine sediments have not been disturbed, and that all the carbonate-accumulating species throughout Quaternary times have behaved in the same way with respect to the oxygen isotopes. Difficulties in the correlation of results from oxygen isotopes with other data may mean that these assumptions are not fully justified.

Time of warmth   Despite such problems in interpreting these data, it is fairly certain that the period of the greatest extent and stability of temperate forest occurred between 5500 and 3000 B.C. during the so-called *post-glacial climatic optimum*, when temperatures reached their highest level since the glaciers retreated. During this period, warmth-loving (*thermophilous*) trees, such as the lime (*Tilia*), reached their greatest range and abundance. The tree-line (the level on mountains above which no trees grow) reached a far higher level than it does today; there is evidence of former forests in the stumps of ancient trees now buried beneath the high-level peats.

The composition of the forest varied as it spread northward. Lime, beech, and hornbeam never established themselves in the far north of the British Isles, because climatic conditions there never became warm enough to support these thermophilous trees. On the other hand, birch and pine flourished in these northern regions where they were subject to less severe competition from the warmth-loving trees. During the climatic optimum, deciduous forest (of oak, elm, alder, lime, and birch) blanketed north-east Europe and the British Isles, probably to an altitude of about 600 m. Much of this high-level forest has since disappeared, due to a combination of human activity and climatic change. The effects of man will be discussed in a later chapter, but the climatic changes that have occurred since 3000 B.C., when the period of climatic stability came to an end, may be briefly noted here.

Although there has been an overall cooling of the climate during the past 5000 years, this has not taken place in a gradual way, but in a series of steps. Beginning about 3000 B.C., a number of quite sudden temperature falls in the Northern Hemisphere had the effect of halting the retreat of glaciers and of increasing the rates of bog growth. One of the most pronounced of these steps, as far as north-west Europe is concerned, occurred about 500 B.C. and caused a sudden increase in the growth rate of bogs over the entire area. In many bogs this has left a permanent mark upon the stratigraphic profile of the bog; a dark, oxidized peat typical of slow-growing bog surfaces is suddenly replaced by the almost undecomposed vegetable matter that typifies a fast-growing bog. The German botanists who first described this phenomenon called it the *Grenzhorizont* (boundary horizon) and the name is still frequently used by palaeobotanists.

Recorded history As soon as man appeared on the scene, he began to leave information about his climate and its changes. Early records are clues rather than precise information, such as the ancient rock drawings of hunting scenes discovered in the Sahara, which indicate that its climate was much less arid at the time they were made than it is today. With the development of writing, accurate records of climatic changes began to be made. For example there are records of pack-ice in the Arctic seas near Iceland in 325 B.C., indicating the very low winter temperatures at that time. During the heyday of the Roman Empire, however, there was a steady improvement in climate, allowing the growth of such crops as grapes (*Vitis vinifera*) and hemp (*Cannabis sativa*) even in such relatively bleak outposts as the British Isles. This warmer, more stable period reached its optimum between A.D. 1000 and 1300, after which it again grew colder.[19]

In 1250 alpine glaciers grew and pack-ice advanced in the Arctic seas to its most southerly position for 10,000 years. In 1315 a series of poor summers began, and these led to crop failures and famine. Climatic deterioration continued and culminated in the "Little Ice Age" of A.D. 1550–1850, during which the glaciers reached their most advanced positions since the end of the Pleistocene glacial epoch. During this time, trees on the central European mountains were unable to grow at their former altitudes due to the increasingly cold conditions.

Since 1700 the climate has become warmer until the middle of the present century, since when it has again grown cooler. After 1940 winters became colder, and since 1950 summers have been cooler and wetter. During historic times all changes in plant and animal distributions have probably been affected to some extent by man, and as far as climatic change is concerned we must be cautious in our interpretation of them.

Because of the complexity of man's influence on his environment it is difficult to tell whether some recent changes in the distribution of certain organisms have been influenced by climate. It is possible that some have, such as that of the lake-dwelling holly-leaved naiad (*Naias marina*), which was fairly widespread in the British Isles during the early post-glacial warm period but is now virtually extinct as a direct result of the overall fall in temperature, surviving only in a few localities in Norfolk. A further example is the lizard orchid (*Himantoglossum hircinum*), a scarce and beautiful plant of south-east England. During the first half of this century it extended its range considerably, but this has contracted again since 1940, possibly as a response to the colder climate since that time. All orchids are extremely sensitive to changes in their environment and many species may have their distribution limited by minor climatic fluctuations.

Causes of glaciation   There are many theories which try to account for the climatic oscillations of the Quaternary Era. Some scientists have suggested that the causes of these climatic changes are primarily extra-terrestrial, whereas others have thought that the answer lies in terrestrial changes, including those brought about by changes in the earth's atmosphere. Some of the theories that have been put forward are outlined below.

The variable zone of dust (probably originating mainly from minute meteoric particles), which surrounds the earth at a height of about ten miles is very likely to affect climatic conditions on earth by blanketing the sun's rays and increasing precipitation of rain and snow by providing nuclei for condensation of water vapour in the atmosphere.

Volcanic activity on earth also contributes to the layer of the dust in the stratosphere and it has been suggested as being important in effecting climatic change, but no correlation has been observed between periods of glaciation and those of marked volcanic activity.

The glacial epochs during the earth's history have always followed periods of mountain-building and this fact has led some geophysicists to suggest that the mountain-building released carbon-dioxide which raised world temperatures by forming an insulating blanket. As the igneous rocks weathered, so carbon-dioxide was absorbed and temperature reduction and glaciation resulted.

The American geologists Maurice Ewing and William Donn have put forward an ingenious explanation to account for the glacial/interglacial cycles.[20] The precipitation necessary for the formation of ice sheets, they postulate, came about as a result of the Arctic Ocean being open and ice-free. This could have occurred if water from the Atlantic had free access to the Arctic, as well it might during warm periods. But as more and more of the world's water became bound up as ice, the level of the seas fell, so that the Arctic Ocean eventually became isolated from the Atlantic by the exposure of a ridge of land extending from north-west Europe to Iceland. Isolated from the warm marine currents, the Arctic Ocean froze and the precipitation in Arctic regions was thus greatly reduced. The glaciers then stopped growing and started to retreat, causing a rise in world sea-levels once more, until a new glacial cycle began. This mechanism could have produced the alternating glacial and interglacial periods, but does not explain how these climatic cycles *started*. Ewing and Donn suggest that continental drift provides the answer: palaeomagnetic evidence indicates that before the Pleistocene the magnetic North Pole was somewhere in or near the North Pacific Ocean, and the magnetic South Pole was in the region of the Southern Ocean. At this time then the Arctic Ocean could not freeze over as ocean currents mixed warm Equatorial water with that of the polar regions. The world climate was then more or less uniform and very mild. The Pleistocene Ice Age began when the North Pole migrated from the North Pacific to a position within the enclosed Arctic Basin, when the process described above began.

There are two main objections to this hypothesis: first, there is not much evidence for such "polar wandering" as late as the end of the Tertiary period, and second, if the source of precipitation was a completely ice-free Arctic Ocean, there surely would have been more glaciation north of the mountains of Alaska and in north-east Asia. Geological studies have indicated that this did not happen; moreover, had these areas not been ice-free, animals and plants from the Old World would not have been able to colonize North America.

We will now turn to those theories which suggest extra-terrestrial causes for the climatic fluctuations of the Quaternary. Some scientists have considered solar disturbances, such as sunspot activity, to be important; such changes could alter meteorological conditions on the earth, but recent research has shown that sunspot activity by itself could not account for the climatic fluctuations. Another theory is that the sun undergoes a rhythm of expansion and contraction and that the intensity of solar radiation varies with this rhythm.

One of the most convincing theories was put forward by a Yugoslav physicist, M. Milankovitch, who calculated the varying amount of solar energy reaching the earth during the past million years. His calculations were based on the facts that the earth's orbit round the sun is elliptical, that the ellipse itself slowly changes its position (or precesses) in space, and that the angle of the earth's axis to the orbit wobbles like a spinning top, taking a definite time (26,000 years) to complete one "wobble". Milankovitch's investigations have produced results showing a beautifully close agreement between times of glaciation and those periods when the earth received least solar radiation. The most feasible explanation of the cause of the climatic ups and down of the Quaternary period seems to be that of Ewing and Donn or that of Milankovitch— or maybe a combination of both.

However convincing the theories outlined above may seem, the question of how the great periods of glaciation occurred has by no means been solved; the mechanism is certainly a complex one with many variables.

We have seen how the movements of continents and climatic changes have affected the distributions of animals and plants throughout the earth's history; now we shall see the effect man has had on the distribution of the other living things with which he shares this planet.

References

1   GOOD R. (1964) *The Geography of the Flowering Plants*, 3rd edn. Longmans, London.

2   DARLINGTON P.J. (1965) *Biogeography of the Southern End of the World*. Harvard University Press, Cambridge, Mass.

3   MULLER J. (1970) Palynological evidence on early differentiation of angiosperms. *Biol. Rev.* **45**, 417–450.

4   DOYLE J.A. (1969) Cretaceous angiosperm pollen of the Atlantic coastal plain and its evolutionary significance. *J. Arnold Arboretum* **50**, 1–35.

5   DARLINGTON P.J. (1957) *Zoogeography: the Geographical Distribution of Animals*. Wiley, New York.

6   COX C.B. (1973) Systematics and plate tectronics in the spread of marsupials. Organisms and Continents Through Time, (ed. N. F. Hughes). *Palaeont. Ass., Spec. Pop. in Palaeont.* **9**, 113–119.

7   PATTERSON B. & PASCUAL R. (1968) The fossil mammal fauna of South America. *Q. Rev. Biol.* **43**, 409–451.

8   COOKE H.B.S. (1968) The fossil mammal fauna of Africa. *Q. Rev. Biol.* **43**, 234–264.

9   AXELROD D.I. & BAILEY H.P. (1969) Paleotemperature analysis of Tertiary floras. *Palaeogeog., Palaeoclimat., Palaeoecol.* **6**, 163–195.

10 WOLFE J.A. (1971) Tertiary climatic fluctuations and methods of analysis of Tertiary floras. *Paleogeog., Palaeoclimat., Palaeoecol.* **9**, 27–57.

11 WEST R.G. (1968) *Pleistocene Geology and Biology*. Longmans, London.

12 DEEVEY E.S. (1949) Living records of the ice age. *Sci. Amer.*, May 1949.

13 WEST R.G. & WILSON D.G. (1966) Cromer forest bed series. *Nature, Lond.* **209**, 497.

14 COOPE G.R. (1965) Fossil insect faunas from late Quaternary deposits in Britain. *Advancement of Science*, March 1965.

15 PENNINGTON W. (1969) *The History of British Vegetation*. English Universities Press, London.

16 GODWIN H. (1956) *The History of the British Flora*. Cambridge University Press, London.

17 FAEGRI K. & IVERSEN J. (1964) *A Textbook of Pollen Analysis*. Munksgaard, Copenhagen.

18 EMILIANI C. (1955) Pleistocene temperatures. *J. Geol.* **63**, 538.

19 LAMB H. (1965) Britain's changing climate. In: *The Biological Significance of Climatic Changes in Britain*. Academic Press, New York and London.

20 EWING M. & DONN W.L. (1958) A theory of ice ages. *Science* **127**, 1159.

CHAPTER 8

THE MARK OF MAN: HIS EARLY DAYS

In Chapter 3 it was explained that all organisms have certain energy and inorganic nutrient requirements and that each species is adapted in such a way that it is able to obtain these requirements within its own particular situation. Each species has thus acquired a niche in the balance of nature which it occupies more efficiently than any other species. In Chapter 4 we saw how the adaptations necessary for the occupation of a niche came into being under the influence of natural selection. Man also has undergone such adaptation during the course of his evolution, as a result of which his ecological niche has changed, not always gradually but sometimes in a series of leaps.

Man's adaptations The early morphological and physiological evolution of man provided him with a high brain capacity, with the ability to walk upon his hind feet, thus enabling his hands to develop manipulative dexterity, and with a high degree of physiological flexibility, allowing him to inhabit a variety of climatic situations. These adaptations, together with a dental structure permitting an omnivorous diet, gave to man the potential to occupy a number of ecological roles. He could feed directly on the fruits, roots and leaves of plants, acting as a herbivore, or he could prey upon other animals. As a predator he suffered several disadvantages for he lacked the powerful jaws, teeth and claws typical of the predator. Also as a biped his speed and manoeuvrability were limited. This would not, of course, prevent him from feeding upon small prey, insects, shellfish, etc., but developments of another kind were necessary for the exploitation of larger prey.

Further adaptation of early man took place in his culture as well as in his body. These adaptations were of two types. In the first place he took to hunting in groups, which made it possible to drive game into situations where they were at a disadvantage. In the second place, the freedom of his hands from locomotory duties enabled him to develop the art of using artificial tools and weapons. This characteristic is almost unique in the animal kingdom. Even the use of unsophisticated implements such as rocks and sticks would have made up for many of his short-comings as a predator.

The ecological niche of early man The fossil history of man is incomplete.[1] We have records of a number of hominid types which are related to one another in a way similar to the branches of a tree—they

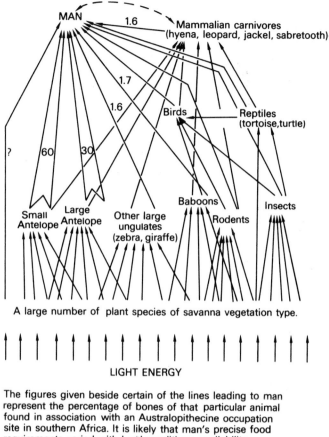

MAN 1.6

Mammalian carnivores
(hyena, leopard, jackel, sabretooth)

1.7

1.6   Birds   Reptiles
(tortoise,turtle)

?   60   30

Small   Large   Other large   Baboons   Insects
Antelope Antelope   ungulates   Rodents
(zebra, giraffe)

A large number of plant species of savanna vegetation type.

LIGHT ENERGY

The figures given beside certain of the lines leading to man
represent the percentage of bones of that particular animal
found in association with an Australopithecine occupation
site in southern Africa. It is likely that man's precise food
requirements varied with local conditions, availability
of game, etc.

— — — — Competitive relationship

———————— Predatory relationship

Figure 66. Foodweb diagram involving Australopithecine primates.

cannot be considered as a linear sequence. Some of the oldest fossil
records of hominid types come from southern Africa and are collectively
termed Australopithecines. From the study of the occupation sites of
these organisms it is possible to reconstruct the position which they
held in relation to other contemporary organisms. Figure 66 is an attempt
to construct a food web of an Australopithecine from the evidence of
bones found in association with his camp sites.[2] The difficulties of
interpreting such sites are great, and the diagram must be a mere
approximation to the truth; for instance, no indication can be given of
the importance of plant produce in this creature's diet.

The predominance of small game in the diet of the Australopithecines undoubtedly reflects their crude hunting techniques and weapons.
The diet of Palaeolithic man[3] (see Fig. 67) has a far greater content of

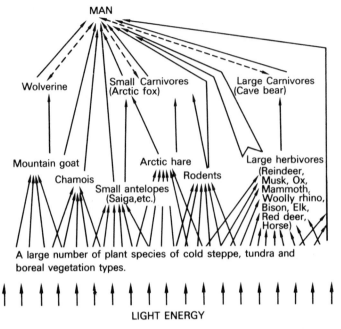

A large number of plant species of cold steppe, tundra and boreal vegetation types.

LIGHT ENERGY

The data from which this web was compiled gives no quantitative information concerning the abundance of the bone of the various animals, except to indicate the predominance of the large herbivore group in the bone assemblage.

— — — — Competitive relationship

———————— Predatory relationship

Figure 67. Foodweb diagram involving *Homo sapiens* at the close of the last glaciation.

large mammals, reflecting the more refined weapons which were available to man almost a million years later. During this period he had also learned how to use fire for cooking and also possibly for driving game. Thus, by the close of the final glaciation man had the capacity to cause widespread modification of his environment. Circumstantial evidence suggests that this is precisely how man began to use his newly found capacities. It was during Palaeolithic times that many species of large herbivores and carnivores (the "megafauna") became extinct.[4] Undoubtedly these creatures were placed under stress by the changing climate (see Chapter 7) and the corresponding changes in vegetation. The poleward extension of the forests, following in the wake of retreating glaciers, invaded the open tundra and steppe which was the habitat of these creatures. However there had been other interglacials which these species had survived as relict populations in the northern extremities of Eurasia and America. One can only surmise that the cultural development of *Homo sapiens* during the final glaciation empowered him to perform the "overkill" which resulted in the extinction of these cumbersome animals, easy prey to man's sophisticated weapons and hunting techniques.

As the glaciers retreated from northern Europe, man followed in their wake. As the vegetation began to develop in its successional stages, responding to the growing warmth of climate, so man took his place in the succession as a hunter and a gatherer.

Close behind the ice followed tundra vegetation, rich in arctic and alpine floras. Bare landscapes were the home of large herds of herbivores and close on the heels of the herbivores came man, occupying the predator trophic level. Such was the way of life of the reindeer-hunting cultures of northern Europe 12,000 years ago.

Further south the birch and pine forests of Europe flourished and spread steadily northwards. Here man developed the skills involved in the hunting of forest animals. To the west of Europe the Romanellian cultures hunted small game such as hare and birds[5] and supplemented their diet with snails and fruit.

Although primitive man drew upon the natural resources of his surroundings, the effects which he had upon the general balance of living things were small compared with his present influence. The reason for this is twofold. In the first place his population density was low, and his environment was therefore easily able to absorb his activities. In the second place his manipulation of the environment had only just begun. He had already learned to use and modify natural objects such as pebbles to produce tools and weapons which, as we have seen, offered man an opportunity to widen the scope of his diet, while the mastery of fire production allowed a wider variety of foods to be eaten and also may have proved useful in hunting.[6] However, despite these cultural advances, man had not yet devised a means of channelling the natural productivity of his environment to his own advantage.

The neolithic cultural revolution It was around 9000 B.C. that a revolutionary cultural change occurred in south-west Asia which was to have a profound effect upon the history of man and of our planet. Here, in the fertile region of Palestine and Syria (Fig. 68) grew the ancestors of our wheat and barley—annual grasses which were expanding northwards as the climate became warmer. *Triticum boeticum* is a wild wheat of this region which may be the ancestor of *T. monococcum*, Einkorn. Both of these have two sets of chromosomes in the nuclei of their cells (diploid). Another wild wheat species, *Triticum dicoccoides*, has four sets of chromosomes in each nucleus (tetraploid) and is therefore probably closely related to a cultivated tetraploid species *T. dicoccum*, Emmer, which has been found amongst Neolithic remains of about 6000 B.C. in Iraq.[7] Most of our modern wheat species are hexaploid, i.e. they have six sets of chromosomes per cell, like *T. aestivum*, the Bread Wheat. No wild hexaploid species of wheat are known and it may be that such species were produced by man as a result of crossing a tetraploid wheat with a diploid grass and then a doubling of chromosomes occurring. Such hexaploid species appeared much later.

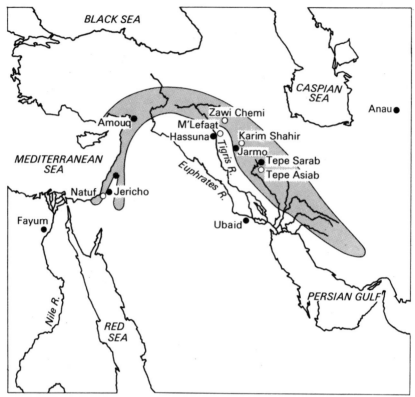

Figure 68. The "fertile" crescent region of the Middle East, the site of the agricultural revolution.

Other wild plants of the Middle East were grown and cultivated by man (see Fig. 69), among them barley, rye, oat, flax, alfalfa, plum and carrot. Further west, in the Mediterranean basin, yet more native plants were domesticated, including pea, lentil, bean and mangel-wurzel. As the revolution continued in other regions of the globe, so the native plants of each region were selected and bred for cultivation: in south-west Asia there were millet, soybean, radish, tea, peach, apricot, orange and lemon; central Asia had spinach, onion, garlic, almond, pear and apple; in India and south-east Asia there were rice, sugar cane, cotton, and banana. Maize, New World cotton, sirol and red pepper were originally found in Mexico and the rest of Central America, while tomato, potato, common tobacco, peanut and pineapple first grew in South America.[8]

In some cases there may have been independent cultivations of the same or similar species in different parts of the world. Thus Emmer may well have originated quite independently in the Middle East and in Ethiopia.

The question of the origins of cultivated cotton has posed many problems.[9] Cultivated cotton has been known for many millennia in Asia and in Central America. The types of cotton growing in these two regions, however, differ much in their genetical constitution. The New World cotton is a tetraploid species of *Gossypium*, having a chromosomal complement which can be represented AADD. The Asiatic species is a diploid with a chromosomal constitution AA only. Wild Asiatic cottons also have the genetic make-up AA, while the wild American diploid species are DD. Thus we have the problem of explaining how the A types of chromosome found their way across the Pacific to the New World.

There are two main theories, the most romantic of which is that plants of A constitution were carried from west to east across the Pacific by early Polynesian sailors—a type of Kon-Tiki theory. The alternative is that the A genome reached America from Asia or Africa before the splitting up of the continents. If this occurred, then wild AA plants might have hybridized with wild DD; this, followed by chromosomal doubling, could have given the AADD tetraploid. It will be interesting to see whether such a wild AA *Gossypium* species still persists, yet undiscovered, in some remote region of Central or South America.

The domestication of certain animals may have preceded that of plants. There is some evidence, for instance, that earlier cultures had domesticated the wolf or, in some North African communities, the jackal. Such animals were probably of considerable use in driving and tracking game and hunting down the wounded prey.

However, it is likely that many of the other animals which became associated with man, such as sheep and goats, were domesticated during the early Neolithic period soon after the first cultivation of plants. These were initially herded for their meat and hides, but would have also been a source of milk, once tame enough to handle. The first traces of domesticated goats come from Palestine around 6000 B.C. These remains are of an animal closely resembling the wild bezoar, *Capra hircus aegagrus*, which still inhabits parts of Asia Minor.[10]

Domesticated sheep appeared at roughly the same time and may have originated from one of the three wild European and Asiatic sheep, or may have resulted from interbreeding among these species. The Soay sheep of the Outer Hebrides almost certainly originated from the mouflon, either the European *Ovis musimon* or the Asiatic *O. orientalis*.

The auroch, *Bos primigenius*, was a frequent inhabitant of the mixed deciduous woodland which was spreading north over Europe during the post-glacial period. In many of the sites where remains of these forests have been preserved, such as in buried peats and submerged areas, the bones of this animal have been found. It was probably first domesticated by the Neolithic farmers of south-east Europe and was taken by them to other areas as their culture spread.

The domestication of plants and animals represents a turning point in the history of life upon this planet. For the first time in the course of evolution, one of the products of evolution was utilizing the principles of evolution to his own advantage. In breeding domesticated organisms, as Darwin readily recognized, man was replacing natural selection with his own volitional selection of those characters which best suited his

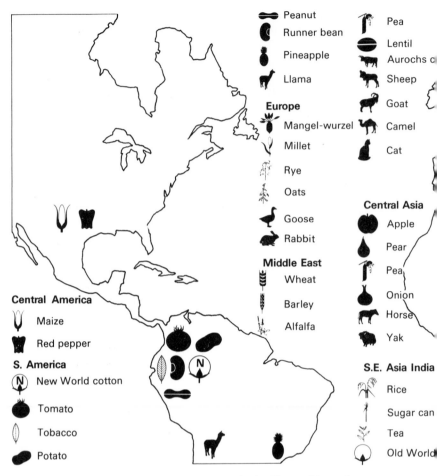

Figure 69. The origins of domesticated plants and animals. The positions of the symbols represents probable sites of initial domestication.

requirements. In this way he became able to control both the course and, to some extent, the speed of the evolutionary process.

Evolution was not the only process which man mastered when he domesticated plants and animals; he also began on the process of controlling his external environment. Up to this time man had been at the mercy of his environment from the point of view of energy supply. Those cultures which relied upon reindeer as their food source were obliged to follow the reindeer herds on their migrations. Those who gathered the fruit of the forest and the earth were utterly dependent upon their environment. Their distributions were those of their staple foods, and with their foods they responded strictly to variation in climate and soil. With the domestication of plants and animals much of this began to change, for man had begun to create an artificial ecosystem around himself over which he would have control and to which he would no longer be subject.

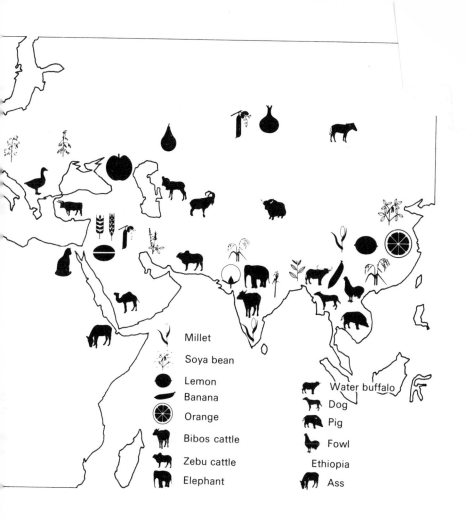

Millet

Soya bean

Lemon

Banana

Orange

Bibos cattle

Zebu cattle

Elephant

Water buffalo

Dog

Pig

Fowl

Ethiopia

Ass

The ecological niche of agricultural man   In the cultivation of cereals and food plants, man developed a primary producer to his own requirements—a plant which would lay up much of the produce of its energy fixation in a grain that was both palatable and easily harvested and stored. In his goats, sheep, cattle and pigs, man developed the second trophic level of his artificial system. These beasts could exploit unprofitable herbage and convert it into energy-rich protein. It may well be that the domestication of these animals began with wild beasts raiding his crops. A semi-parasitic association may thus have been set up which benefited the beast and which man was later able to convert into a mutually beneficial symbiosis. In all cases the ranges of both animals and plants were increased as a result of their association with man: their pests, parasites and predators were attacked by him; those creatures

159

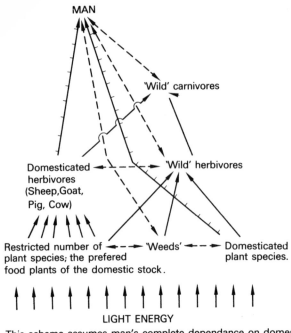

MAN

'Wild' carnivores

Domesticated herbivores (Sheep,Goat, Pig, Cow)

'Wild' herbivores

Restricted number of plant species; the prefered food plants of the domestic stock.

'Weeds'

Domesticated plant species.

LIGHT ENERGY

This scheme assumes man's complete dependance on domestic animals and plants. This extreme situation will have developed gradually through agricultural history.

———————— Predatory relationship

⌐┴─┴─┴─┴─┘ Predatory/mutualistic relationship

– – – – – – – Competitive relationship

Figure 70. Foodweb of agricultural man.

which competed with them were eliminated by him, and they benefited from his increasing control over soil and season. These effects are summarized in the foodweb diagram, Figure 70.

In fact, many of the plants and animals domesticated by man became more and more dependent upon him. As a result of this, many of their adaptations to survival in the wild became less essential for their continued existence and were gradually lost from their genetic constitution. For instance, the long horns of the auroch gradually disappeared and by 3000 B.C. the short-horned race, *Bos longifrons*, was associated with Neolithic people in Europe. This change was the result of constant selection on the part of man. Again, improvement in the palatability of the domesticated carrot involved a selection for a fleshy, soft tap root structure. This involved a higher water content and hence a loss in frost resistance, together with the greater risk of being consumed by other animals. So with domestication and dependence came modification of the plants and animals concerned.

Post-glacial dispersion of man Man's physiological and cultural adaptability, together with his omnivorous diet, has been one of his greatest selective advantages in the struggle for survival. Hence his instinct to spread and to occupy all available regions, however unpromising, has constantly resulted in migration and population diffusion. On occasions physical barriers were converted into aids for dispersal as a result of his cultural ingenuity. Water, which presented an insurmountable obstacle to the movement of so many organisms, offered a means of transport to the man with a knowledge of navigation. As nautical arts developed, so even the sea ceased to be a cultural barrier and became instead a link between groups living on coastal plains, leading to an interchange of ideas and crafts and thus an acceleration in the development of man's mastery over his environment. Communications were often so much more satisfactory by water than via a densely-forested interior that a coastal residential stability proved a most efficient way of life for the people of Europe during the post-glacial climatic optimum of 5000–3000 B.C. The forests had by this time reached their maximum extent and covered most of the continent, the deciduous trees giving way to a predominance of conifers in northern Britain, Scandinavia, Finland and Russia.

In north-west Europe, rising sea levels during this period had cut Britain off from the continent of Europe, and Ireland from the remainder of Britain. This part of Europe still remained in a pre-agricultural condition—the Neolithic revolution had not yet taken place. The village farming system of south-east Europe, however, combined with seafaring arts proved so successful that by the year 3000 B.C. it had swept up the western seaboard of Europe and was invading hinterlands wherever there was access from the sea and where soils could support their herds and crops. So the influence of man upon his environment first began to show itself as he reconstructed his surrounding ecosystems to redirect the available energy towards his own support.

Man's forest environment The natural stable or climax ecosystem which predominated in western Europe when Neolithic culture arrived, was deciduous summer forest—a highly diverse and complex ecosystem containing a large number of plant and animal species. The vegetation is spatially complex, different species supporting their leaves at various levels above the ground. These levels can be conveniently divided into tree canopy, shrub canopy, herb layer and moss layer, but this is an oversimplification of the highly complex stratified structure of deciduous forest. Such stratification results in an efficient light-trapping system, very little light being allowed to penetrate to the soil and therefore lost to the plants. There is also a temporal structure to woodland, for not all plant species emerge or put out their leaves at the same time. For example, many of the herbaceous plants of deciduous woodland grow most rapidly during the early spring whilst light intensities are high, before the shrub and tree canopies open. Such complex structure leads to a large number of niches for animal species, hence species diversity is high and food webs are extremely complex. In general it is agreed amongst ecologists that such complexity in natural systems is often associated with considerable stability.

The complexity of woodland is found not only in its energy relationships, but also in its nutrient balance. Climax forest is in a state of equilibrium with its physical environment, which means that the circulation of nutrient materials between the various components of the ecosystem must be critically balanced. One can consider each component as a reservoir for nutrients, some containing larger supplies than others, according to their bulk and the concentration of nutrients within them (see Fig. 71). The trees of oak forest, for example, largely because of their bulk and their high nutrient requirements, represent an

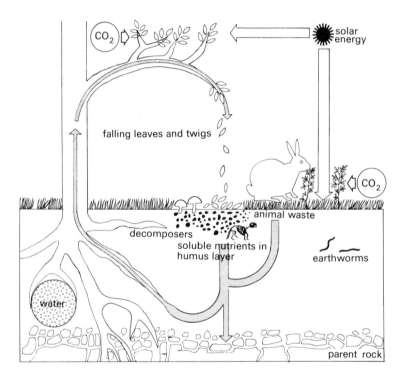

Figure 71. Schematic nutrient cycle in a forest ecosystem.

important reservoir of nutrients for the system.[11] As individual trees die, so they pass to the decomposer organisms, and the nutrients they contain are eventually "mineralized" or returned to the soil in inorganic form. Thus nutrients are released gradually from the "tree reservoir" for re-use. The soil itself varies in its properties as a reservoir. Some soils are efficient, being rich in the clay minerals which bind the nutrients in the soil and prevent them from being washed away by rain. Other soils can be thought of as "leaky" reservoirs having a sandy, well-drained constitution with a low clay content. Such soils are inefficient as reservoirs and on their own soon become depleted of nutrients.

When man replaces such forest with an arable crop, he reduces both the complexity and the stability of the system. He reduces the species diversity of the system—often, in the case of plants, to a single species.

The changes in food-web relationships in man-made and man-managed ecosystems have already been mentioned. In arable farming he removes all other consumer organisms and acts as a herbivore. In pastoral farming he retains a single herbivore species (e.g. sheep or cow) and occupies the position of sole carnivore. This brings man into a state of active competition with all plants apart from his favoured ones, all herbivores apart from those he has adopted and all carnivores feeding upon his domesticated herbivores. Food webs are thus made simple, short and linear (see Chapter 5).

Removal of trees, or burning off vegetation to clear areas constitutes the loss of a major nutrient reservoir. Where burning is used, the nutrients are liberated, but if the soil has a low capacity for nutrient retention, then the nutrients do not remain long in the soil but are leached away. A crop is then sown which draws upon the soil's remaining nutrient reserves, but on reaching maturity it is harvested with no return of nutrients to the soil. In such a situation the only replenishment of the nutrient capital of the ecosystem comes from the weathering of parent rocks and to a certain extent from the atmosphere, both of which may well prove too slow to cope with the constant drain on resources caused by an annual harvest. If a grazed grassland system replaces the woodland, then simplification of the nutrient cycle still occurs. Grasses do not compose as great a reservoir for nutrients as did the trees, hence once again the soil is placed under stress as livestock is regularly harvested from the system.

Grazing and soil stability   Overgrazing can also cause problems of soil instability. Soil under grassland is maintained in a stable state by the fibrous roots of the grasses and the cover provided by the above-ground parts. Grasses can withstand moderate grazing by virtue of the fact that leaves grow from the base rather than the tip, so removal of tips does not stop growth but may even promote it by encouraging light penetration of the sward. However, a minimum amount of leaf and stem tissue must be left after grazing to allow regeneration and to prevent soil being blown away. If an area of grassland is overstocked with herbivores, then even this minimum is removed and the soil becomes unstable. This occurred in the southern Great Plains area of the United States, where an unpredictable climate aggravated a situation in which overgrazing and wheat farming were practised. Drought in the latter part of the nineteenth century resulted in the abandonment of much of the grazing, but with the rains at the turn of the century came the ploughing of the area and the growth of the wheat industry. Within thirty years the effects of cropping and laying bare the soil, together with periodic severe droughts, had made themselves felt, and the soils of the southern Great Plains became unstable. By this time the soil had become so impoverished that it was impossible for native grasses to recolonize during drought periods. The soils became mobile and the area received the name "Dust Bowl".

Soil devoid of vegetation for any length of time is liable to lose its structure. When denied a constant input of organic matter, the crumb structure of soil disintegrates as the microflora and fauna becomes poorer. Such unconsolidated soil is subject to wind erosion and this is precisely what transpired in the Great Plains.

Man's effects upon natural ecosystems can be summarized as:

1 Reduction in species diversity, often to a single species population.
2 Reduction in the complexity of food webs.
3 Removal of many of the "reservoirs" of nutrient material from the ecosystem, some of which may be of major importance (e.g. trees).
4 Disruption of the course of nutrient cycling, (a) by harvesting without a compensating return of minerals and organic material to the soil, or (b) in more recent times, by flushing an excess of nutrients (waste products and pollutants) into ecosystems.

Ecosystem simplification of this type carries with it many dangers. In the first place, a single species population, such as a field of wheat or a herd of sheep, offers great opportunity for the development, spread and ravages of disease, pests and parasites. Simple systems are often unstable ones and may also suffer from erosion unless carefully managed.

Human distribution patterns   We have seen in Chapters 1 and 2 that both plants and animals are distributed according to their numerous climatic, physical and chemical requirements. Man is no exception to this basic principle, but he has the good fortune to possess certain attributes which have proved most successful. In the first place man has an extremely adaptable physiological make-up which has permitted his survival even under adverse environmental conditions. In other words his basic ecological requirements are broad. In the second place his high brain power and manipulative skills have combined to give him a degree of control over his immediate environment which has enabled him to adapt it to his needs. He has constructed clothing and housing to his requirements which can be considered as the development of a micro-ecosystem around his body in which climatic variables are under relatively strict control.

A chink in the armour, however, lies in man's nutritional require-ments. Agricultural man could occupy only those regions suitable for the survival and healthy growth of his cultivars and, since the environ-mental tolerances of these organisms were normally narrower than those of man himself, man's distribution patterns often reflected the requirements of his cultivated species. This problem was overcome to some extent, however, by having available a number of food plants and animals and also by selecting strains of these organisms which were resistant to environmental stress whether climatic or biotic (e.g. disease).

An interesting demonstration of the factors affecting the distribution of prehistoric human communities of various cultures is afforded by the west of Britain, and in particular Wales. The hunting cultures at the close of the glacial period were dependent upon reindeer and mammoth as a food source, and their distribution probably reflects that of the herds of these animals. As the forest spread and became dense, so human groups were forced to dwell upon the coastal fringes. Neolithic culture probably arrived from the sea. Boats containing these people and their domesticated plants and animals are thought to have made their way northwards up the western seaboard of Europe.[13] The rocky shores and promontories with sheltered harbours would have been most attractive to such people. Here the coastal soils are likely to have been shallow and the forest cover light. Such areas were easily cleared and used for grazing and cereal growing. The spread of such people inland would have been hampered by dense forest growing upon heavy soils which could not have provided easy conditions for cultivation.

If one examines a map of the distribution of Neolithic monuments, which are perhaps a good indication of where permanent communities of these people existed, then it can be seen that the distribution is most uneven (see Fig. 72a). Concentration of population, as might be

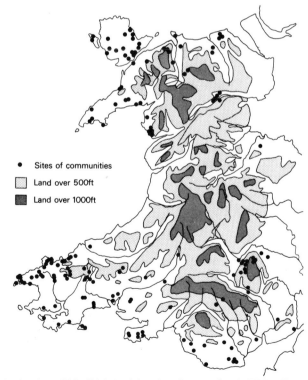

Figure 72a. Distribution of Neolithic burial and settlement sites in Wales. Data from W. F. Grimes (1965).

165

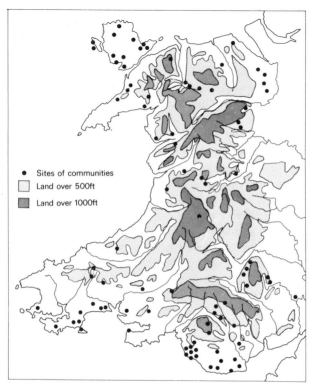

Sites of communities

☐ Land over 500ft

▨ Land over 1000ft

Figure 72b. Distribution of early Bronze Age burial and settlement sites in Wales. Data from H. N. Savory (1965).

expected, is most dense in the coastal areas where forest clearance, grazing and cultivation would have been easiest.[14] Inland, only the Black Mountains region of Wales bears much evidence of having been occupied. Perhaps the sandstone rocks of this area produced soils which were better drained than the shales of central Wales and were therefore easier to cultivate. Here, then, is a human population being limited by soil factors which operate via his domesticated animals and plants.

The point is further demonstrated if we look at a map of Bronze Age remains in Wales (Fig. 72b); it will be seen that most of the highland areas were now occupied.[15] The explanation of such a distribution is explained by two facts, (a) the Bronze Age folk were nomadic pastoralists, and (b) in the upland areas the once dense forest had given way to a peat-forming vegetation which would have provided good rough grazing.[16] These vegetational changes had probably come about as a result of the changing climate, which was becoming increasingly wet and cool. So here again we see man's distribution being determined by the requirements of his domesticated creatures as recently as 3000 years ago.

As his culture has become more complex, so man has attained a higher degree of insulation from environmental factors. Quite early in his history a division of labour enabled him to be free from the necessity

166

of living close to sources of raw materials for tools. The axe industries with their complex marketing systems provide ample evidence of this, axes being manufactured where raw materials were present and then exported, often over hundreds of miles to where a demand was present. Later, a similar system was to be employed in agriculture and food production, which gave great impetus to the process of urbanization.

Development of many varieties of domesticated organisms covering a greater range of ecological tolerances and greater disease resistance also provided for a greater independence of environmental factors in food production. But despite these advances, much of the world's surface, the deserts, the seas, the tundra, etc., remain almost entirely unexploited. The need is to find either a suitable crop plant or animal, or to modify the environment to make it acceptable to one of our present cultivars.

Planned modification of the environment, however, has not been a prominent feature of human history. Changes in man's surroundings have almost always been the result of casual exploitation without any concern for long-term consequences. The modern environmental crisis, its effects upon the animals and plants of the world and its implications for the future of mankind will be considered in the final chapter.

References

1 BOUGHEY A.S. (1971) *Man and the Environment*. Macmillan, London.
2 HOWELL F.C. (1966) In: Caldwell J.R. (ed.), *New Roads to Yesterday*. Thames & Hudson, London.
3 DE SONNEVILLE-BORDES D. (1966) In: Caldwell J.R., (ed.), *New Roads to Yesterday*. Thames & Hudson, London.
4 REED C.A. (1970) Extinction of mammalian megafauna in the Old World late Quaternary. *Bioscience* **20,** 284.
5 WATERBOLK G. (1968) Food production in prehistoric Europe. *Science* **162,** 109.
6 SMITH A.G. (1970) In: Walker D. & West R.G. (eds.), *Studies in the Vegetational History of the British Isles*. Cambridge.
7 COLE S. (1961) *The Neolithic Revolution*. British Museum (Natural History), London.
8 WRIGHT H.E. (1970) Environmental changes and the origin of agriculture in the Near East. *Bioscience* **20,** 210.
9 PICKERSGILL B. & BUNTING A.H. (1969) Cultivated plants and the Kon-Tiki theory. *Nature, Lond.* **222,** 225.
10 ZEUNER F.E. (1954) In: Singer C. *et al.* (eds.), *A History of Technology*. Oxford.
11 OVINGTON J.D. (1965) *Woodlands*. English Universities Press, London.
12 DASMANN R.F. (1968) *Environmental Conservation*. Wiley, New York.
13 CLARK G. & PIGGOTT S. (1965) *Prehistoric Societies*. Hutchinson, London.
14 GRIMES W.F. (1965) In: Foster I.L. & Daniel G. (eds.), *Prehistoric and Early Wales*. Routledge & Kegan Paul, London.
15 SAVORY H.N. (1965) In: Foster I.L. & Daniel G. (eds.), *Prehistoric and Early Wales*. Routledge & Kegan Paul, London.
16 MOORE P.D. (1968) Human influence upon vegetational history in North Cardiganshire. *Nature, Lond.* **217,** 1006.

CHAPTER 9

THE MARK OF MAN; MODERN PROBLEMS

In the last chapter the early history of man was traced with especial reference to the response of the environment to his increasing population and consequent demands for more space and food. Changes in the organization and structure of human society during the past few centuries have greatly accelerated these processes and have resulted in the degradation of the environment.

These processes have included:

1 Industrialization, as a result of which man has made new demands upon the non-renewable resources of the planet.

2 Urbanization—the concentration of population into certain areas, mainly the centres of industrial activity.

3 Intensive agriculture—a response to high demand from urban populations aggravated by limited space.

1 *Industrialization*   Although it is difficult to put a precise date to the "Industrial Revolution" in northern Europe, it is generally accepted that, during the latter part of the eighteenth century, industrial development received great stimulus fron the pressures of scientific invention and rising population.[1] At this time industry became a critical factor both in social and in environmental spheres. Early industrial development was concentrated in areas where the necessary raw materials (iron, coal, limestone, other metals, etc.) could easily be obtained.

Coal represented a source of energy for industry, providing a means for the extraction of iron from its ores. Despite the need for mining to supply coal, it proved far more valuable than charcoal as an energy source. However, certain ecological problems were associated with the use of "fossil fuels" as an energy source, many of which are still not widely appreciated. In the first place, the use of fossil fuels involved the taking of carbon from the ground where it is not available to living organisms, and its conversion into atmospheric CO_2 thus accelerating the cycling of carbon on a global scale and raising atmospheric CO_2 levels. Even in Hawaii, far from the world's major industrial areas, this effect has been noticeable over the last ten years, during which atmospheric CO_2 levels have risen by 10 ppm.[2] The possible climatic implications of such a change have already been discussed in Chapter 7. In addition there is evidence that the total green plant material in the world has risen in response to increased atmospheric CO_2

It is impossible to predict the long-term effects of this consequence of industrial activity. Carbon dioxide is not the only gaseous product resulting from the combustion of coal and oil. All living matter contains quantities of sulphur and, since coal and oil are formed initially from dead organic material, it also has a sulphur content. On combustion this is converted into sulphur dioxide which enters the atmosphere. Unlike CO_2, SO_2 appears to be positively harmful to living organisms, though they vary greatly in their sensitivity to the compound. In particular epiphytic lichens appear to be extremely sensitive to SO_2 (see Fig. 73). These primitive plants, living some distance off the ground by using trees for support, absorb their minerals from rainwater. They do this

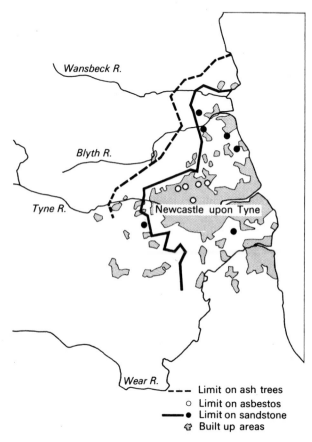

Figure 73. Map showing the limits of distribution of certain lichen species in the Newcastle upon Tyne area. The solid line shows the limit of *Parmelia saxatilis* on sandstone walls, and the broken line shows the limit of more sensitive lichens, e.g. *Evernia prunastre* on ash trees. Dots indicate isolated occurrences. Data from O. L. Gilbert.[8]

over their entire surface rather than by any specialized organs. Water with high concentrations of SO_2 rapidly kills such plants, as a result of which industrial regions are very poor in epiphytic lichens and have been termed "lichen deserts". The effects of SO_2 on animals including man are also unpleasant and can lead to respiratory complaints.

Dust and smoke particles from fuel combustion are more easily detectable than SO_2 and have thus come to the attention of the public sooner. Undoubtedly pollution of this sort has been responsible for much human suffering in the form of respiratory diseases, both directly and by assisting the formation of smog. Fortunately this has decreased in recent years as a result of the use of smokeless fuel for domestic heating.

Mining for coal, iron and other raw materials has itself produced ecological problems, as a result both of extraction methods and of the disposal of waste products. Open cast mining processes and the dumping of slag is a severe problem because the resulting ground frequently proves unsuitable for colonization by vegetation. Because of the toxicity of waste material, large areas of derelict land are being produced by industrial societies (Fig. 74). In the past, little attention has been given to this wasteful process, mainly because society itself has been divided by

Figure 74. The Lower Swansea Valley. Industrial dereliction following mining.

industry. Industrial working classes have been forced by financial circumstances to reside amongst industrial spoil, close to their place of work, whereas politically influential administrative classes have been able to live in better conditions further away.

Industrial spoil has not therefore provided aesthetic problems to those in a position to influence the situation. Neither has it had any financial effect upon industry until relatively recent times, when land has become increasingly expensive and the comfort of the working classes has been recognized as a worthwhile investment. Reclamation of land made derelict by industry has at last begun. At the end of 1964, 99,000 acres of land in England and Wales were classified as derelict, of which 23,000 acres were spoil heaps.[4] The vast bulk of this land is concentrated in the areas where industry had been centred, i.e. northeast England, Lancashire, the west Midlands and South Wales. It has been estimated that at least £30 million is needed to restore this land. This is the cost of an aesthetically pleasing environment, but this must be regarded not as a luxury but as a necessity.

A final point with respect to the needs of industry for raw materials is that recognition must be given to the fact that these materials are non-renewable. The quantities of energy-rich fuels and useful metals upon this planet are severely limited and industrial man can maintain himself in the long term only if he adopts two basic ecological principles. These are, (i) the recycling of all waste materials, and (ii) his ultimate reliance upon an inexhaustible supply of energy, the sun. If these two principles are applied, then human populations can expand and develop until limited by that factor which is in shortest supply. This could be living space, or possibly the supply of a mineral such as phosphorus for food production.

2 *Urbanization* The development of the industrialized society greatly accelerated the process of urbanization, a process which began as soon as division of labour was possible with the development of domestication. Industry provided a demand for manpower, but required that it should be concentrated in the areas where raw materials were available. Such concentrations of humanity into industrial areas produces both social and environmental problems.

Socially, the effects of crowding upon the vast majority of the population has produced a need for space during leisure periods. This has led to an exodus of urban populations to surrounding rural areas for relaxation. Thus the ecology of industrial man forms an important consideration for anyone concerned with land-use planning and environmental conservation. An additional environmental problem arising from urbanization is the disposal of domestic waste and sewage. When human populations were low, the rivers proved adequate for the dispersal of organic waste, and microbial populations could cope with its breakdown. As human populations increased, so did organic waste until the process of microbial degradation in rivers required so much of the water's dissolved oxygen that oxygen shortage resulted. Lack of oxygen soon leads to the death and decay of all living organisms in the water, and a dead river results. Apart from the pollution resulting from such waste disposal, the release of sewage into rivers represents a loss of nutrients such as nitrate and phosphate into the sea.

Domestic and industrial uses of detergents add to the problems of water pollution. Concentrations as low as 0·1 ppm. of detergent in a river may halve the rate at which it is capable of absorbing oxygen, and thus aggravate the oxygen depletion problem.[5]

3 *Intensive agriculture* In the last chapter, some of the problems resulting from man's development of agriculture were considered. Such problems became more acute with the industrial revolution and subsequent urbanization, because of the increasing burden placed upon agricultural land to supply the nutritional requirements of the towns. These demands made it difficult and uneconomic to continue with sound principles of crop rotation—such as periods of legume growth to replace lost nitrates, and fallow periods to restore soil fertility and structure and to reduce the populations of soil pathogens which rapidly build up if a single crop species is grown repeatedly. The desire to produce a larger and more frequent harvest thus necessitated the use of artificial fertilizers and pesticides.

When chemical fertilizers are applied to a soil, many of the elements contained in them are retained by the soil, having formed a complex with the clay and colloidal humus fractions. However, certain ions are not so retained. Among them is nitrate, an important constituent of most fertilizers, which is very soluble and is rapidly leached out of the soil into rivers and lakes. Here it continues to have a stimulatory effect upon plant growth, and soon the native freshwater algae respond by growing more rapidly. Chemical enrichment of water causing this increased productivity is termed eutrophication. Unfortunately the outcome is ultimately harmful, because the plants die and decompose so that oxygen levels in the water fall until life becomes impossible. The results of this form of pollution can be extremely unpleasant, especially when it occurs on the scale recently observed in Lake Erie, Canada, where the shores have become deeply covered with decaying organic matter.

Just as fertilizers have become necessary to maintain or enhance the productivity of soils, so pesticides have been increasingly used to reduce the quantity of the crop passing into food chains other than those leading to man. It is now realized that many of these compounds have effects which go far beyond the original intention in terms of wildlife destruction. Seeds dressed with fungicides have caused high levels of mortality in those birds which eat seeds; compounds such as dieldrin when used in sheep dips have caused deaths among carrion feeders such as the golden eagle in the Western Islands of Scotland. The use of DDT and other chlorinated hydrocarbons as insecticides is now recognized as having far-reaching side effects. These persistent compounds are stored in the fatty tissues of animals and thus find their way into many food chains, often accumulating to toxic levels in long-lived predators. Man himself is accumulating these compounds in his body—a fact which may bring home to him the need to understand the consequences of his actions before he undertakes them on global scales.

Man's cultural development of the past few centuries has been accompanied by vast increases in the world's population (Fig. 75). The combination of these two processes has meant that there are probably no areas remaining on earth which have not been affected by man to some degree. Even Antarctic penguins have been found to

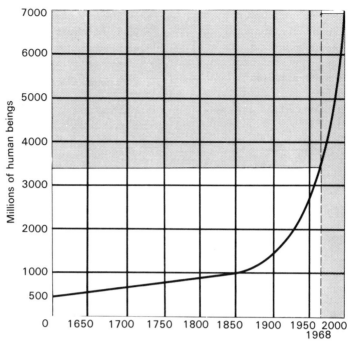

Figure 75. Human population growth curve over the past 400 years.

contain DDT residues. In any consideration of plant and animal distributions, therefore, it is necessary to take into account man's relationship to the organism being studied.

Man and other organisms   There are five main types of relationship which man has developed with other organisms.

1   Organisms providing food.

2   Organisms which have been domesticated and which have therefore entered a symbiotic relationship with man, their success being completely dependent upon his success.

3   Organisms which occupy the same or a similar ecological niche to man or his domesticated symbionts, i.e. competitor organisms.

4   Organisms which occupy the ecological niches which man produced in the course of his activities: the plants and animals which follow in man's wake.

5   Organisms affected by habitat modification and destruction.

1   *Organisms providing food*   In the early days of man's existence, when human populations were low, organisms which were used as food were, on the whole, in little danger of extinction as a result. As population levels rose, however, food animals which provided easy prey, especially the large and clumsy types, came under pressure. Undoubtedly the extinction of many of the large herbivores which were

173

contemporaries of early man can be regarded as due to "overkill". Present-day effects of this kind are fortunately rare, but there is danger of overkill in the fishing industry and even greater danger in the whaling industry where, once again, the victim is hampered by large size and slow breeding rates.

2 *Domesticated organisms* Some consideration has been given to these already. The development of the idea of domestication and of its use in practice was almost certainly a gradual process involving initially what amounts to the management of natural ecosystems. The intelligent hunter is concerned about the welfare of the quarry species and will exhibit this concern by manipulating the environment to favour its survival and population increase. A set of intelligent ethics are formed such as the protection of young animals or pregnant females; a "closed season" may even be introduced to allow breeding. Predators of the quarry species may be eliminated, and plants which it feeds upon may be encouraged, e.g. by burning heather to encourage fresh, young growth for grouse. Man's application of these management principles to hunting soon leads to a state of symbiosis arising, with a mutual dependence developing between hunter and hunted.

Many domesticated organisms, then, originally belonged to class 1, but because of their favourable features were able to enter a closer association with man, and thus their interests and survival were assured. Many organisms which have not attained a fully domesticated status have nevertheless benefited in terms of distribution and population density as a result of their use by man as a source of food. In North America and Europe, game birds provide a good example of this situation. Both native species (grouse) and exotic ones (pheasant) have been carefully managed, protected and harvested by man. Rabbits have benefited in a similar way, despite their tendency to compete with man for his own crops.

3 *Competitor organisms* When man lived as a hunter/gatherer, he competed for food with other carnivores and omnivores. As his population increased and his weapons improved, so the competition became more severe, and reduction in the populations of large carnivores was an inevitable result.[6] Competition also became more intense with the coming of domestication, since this often provided the competitor with an easy meal. For example, the bullfinch feeds extensively upon the flower buds of fruit trees as well as upon the fruits themselves. When man domesticated these trees and grew them in orchards, he provided the bullfinch with an opportunity for population expansion. The resulting competition is still a major problem in southern England.

Domestication also increased the variety of species in competition with men, since all organisms which adversely affect the growth and reproduction of his domesticated organisms must find themselves at enmity with man. His policy has generally been one of elimination of all potential competitors.

4 *Organisms following in man's wake* Some species of plant and animal have been favoured by man's activities and by his spread, having been able to occupy vacant niches in the ecosystems man has created for himself. For example, in the agricultural ecosystem there are many interesting examples of the ways in which wild plants have become adapted for survival. *Camelina sativa* (gold of pleasure) is one example; this plant is a native of eastern Europe, and has probably adopted the role of a flax-field weed ever since *Linum* (flax) was first cultivated. However, man spent much energy in his efforts to maintain a single-species population in his flax fields, and developed the process of winnowing the seeds to eliminate aliens. In this process *Linum* seeds are separated from chaff and smaller seeds by their low surface/weight ratio, whereby they fall rapidly to the ground when thrown in the air, whilst lighter material is blown away. *Camelina sativa*, in response to such selection for a particular seed size, has evolved a variety with seeds of very similar properties to those of *Linum*, which cannot be separated by winnowing. This variety was thus assured of survival and extended range at the hand of man until more effective screening methods were developed.

Not all weeds, however, have been distributed with man's crops from the origins of cultivation. Many were already present, in small isolated populations, when man first began his agricultural practices in a new area. In Britain, for example, many of the plants growing in the open tundra conditions at the close of the glacial period were able to survive through post-glacial times in unstable regions where forest could not establish itself, e.g. on cliffs, sand-dunes, eroded river cuttings, and on high mountains with shallow soils. Such plants could not tolerate the heavy shade of forest, and were driven from most regions of Britain. With the arrival of man and his clearance of woodland, new open habitats were created which once again favoured these relict plants which demanded high light intensities for their growth, and they were able to spread and flourish as man cleared wider and wider areas. Among such plants were mugwort (*Artemisia vulgaris*) and the may-weeds (*Matricaria spp*). It is from the remains of pollen of such species as these, preserved in peat and lake deposits, that the first traces of human interference with vegetation can be detected, and by the relative quantities of such pollen grains, the extent of forest clearance during any particular period can be estimated.

Many animals have also been able to take advantage of the artificial ecosystems which man has provided. Some species which were once cliff dwellers, such as the starling and the rock dove, have established themselves in cities throughout the world, feeding in parks and gardens and returning to roost upon the concrete "cliffs" of city centres. Here are examples of organisms occupying specialized ecological niches at low population densities, which have developed the necessary flexibility to invade and occupy the new niches which man has provided.

An interesting development of the last century has been the occupation of a new niche by the black-headed gull. This essentially sea bird has now become a familiar sight in British arable countryside often over a hundred miles from the sea. Here it scavenges in fields with members of the Corvidae (crows) and is also becoming increasingly common upon urban rubbish tips. Until very recently these inland parties of gulls have been almost entirely restricted to winter, but there are now many sites, particularly around large man-made reservoirs, where the black-headed gull breeds inland. Its occupation of this new man-made niche is thus becoming more complete.

5 *Organisms affected by habitat modification* In certain cases man's disruption of the natural habitat has brought together species which are normally widely separated by their differing ecological requirements. For example, the destruction of woodland and its replacement by meadow in southern Britain has brought together two closely related species of the genus *Primula*. *Primula vulgaris* is the primrose, a yellow flower which is usually found growing in deep shade in woodlands. Its flowers are borne on long individual stalks arising from the centre of a rosette of leaves. The cowslip (*P. veris*) on the other hand, is a closely related species growing in open meadows. It has yellowish orange flowers on short stalks which are all borne at the apex of a leafless floral axis or scape surrounded by a basal rosette of leaves. Under natural conditions, that is in the absence of man, it is most unlikely that these two plants should ever be found together because of their different environmental requirements. However, man has destroyed this natural barrier by clearing woodland and replacing it with fragmentary copses and hedgerows. In this way the two species have been brought into contact and have cross-bred to give a fertile hybrid with characters intermediate between the two parental species. This hybrid has the scape of *P. veris* on top of which are borne yellow flowers with long individual stalks, like those of *P. vulgaris*.

Often, however, disruption and fragmentation of habitats has had even more serious effects upon their occupants. Some species, particularly large predators, require an extensive area of specialized habitat in which to hunt and breed. Fragmentation of such habitats has often had devastating effects upon the population levels of such species. A good example of this is afforded by the marsh harrier, a large raptorial bird of reedbeds and fens. Besides needing a wide area of reedswamp for hunting, this bird will not nest except in the isolation provided by extensive reedbeds. Because of its demand for space, this bird has suffered badly as a result of drainage of wetlands, both in Europe and North America. Its numbers continue to decline, only two pairs succeeding in raising fledgelings in 1970, and it is soon likely to be lost as a breeding species in the British Isles.

Britain was once almost completely covered with woodland, yet now, primarily due to man's activity, only $7\frac{1}{2}$ per cent of its land is wooded. This represents one acre for every 13 inhabitants.[4] In Germany, another industrialized country, there is one acre of woodland to every eight persons and in the United States there are four acres of woodland to each person. In general, those countries which became industrialized before the exploitation of fossil fuel have fared worse as far as woodland depletion is concerned. However, intensive agriculture has also taken its toll of land once wooded, especially in highly populated countries. Twenty-nine million acres of Britain are under crops and grass out of a total land area of 56 million acres. Urban areas constitute 4·5 million acres. Under such conditions Britain is forced to import 90% of its timber requirement. The industralized and agriculturalized nations are thus placing a heavy demand upon the woodlands of the less developed nations. During the last few years 40 per cent of Brazil's forests have been felled.

Biological conservation   One may now ask the question, how can man's relationship to his environment be improved? Man requires industry and agriculture to maintain his way of life; he therefore needs to tap the resources of the planet. The handling of this process must be based upon a sound knowledge of the consequence of any particular line of action, and a prerequisite for this is an understanding of the way in which natural systems operate. The sciences of ecology and biogeography must therefore form the basis of practical conservation.

We have seen that the working of natural ecosystems depends upon a flow of energy through that system and a cycling of matter within it. The aim of conservation must therefore be to maintain and control these two processes. Where energy and matter are to be harvested from a system, there must be an equivalent return of the latter. Populations of organisms within ecosystems are controlled by a variety of factors including food supply, physical conditions, predation, competition, disease, etc. The maintenance of general population balance whilst harvesting certain animals or plants from an ecosystem is a delicate process and requires much knowledge of the dynamics of the populations involved.

The development of "Biological Control", i.e. the application of biological principles in the control of the pest organisms, could make it possible to avoid the use of chemical pesticides with all their harmful side-effects. Biological control systems however must be based upon a thorough knowledge of the population dynamics of the "pest", as early attempts at introducing a pest's predator or parasite have shown. Considerable initial research is therefore required, but some notable successes have already been achieved.

Perhaps the most famous usage of biological control mechanism is the campaign against the prickly pear cactus in Australia, which took place between 1920 and 1940.[7] The cactus had been introduced from America and behaved as a very vigorous weed in Australia where it rendered much valuable grazing land quite useless. Chemical treatment was uneconomic, some 60 million acres being affected. It was decided to introduce an insect enemy of the prickly pear from America, but the choice of pest had to be made with considerable care, lest the insect should prove to be a pest of other plants, possibly economic crop plants. Eventually the moth *Cactoblastis cactorum*, a native of the Argentine, was chosen. This insect has larvae which tunnel in the stems of prickly pear, reducing it to a rotting mass. *Cactoblastis* was reared in captivity and eggs were distributed throughout the affected areas. Between 1920 and 1930, 3000 million eggs had been released. The insect was able to maintain itself in the wild and soon there was no further need of introduction. The initial effects were dramatic, as thousands of acres of prickly pear succumbed to the moth larvae. However, as the pear disappeared, so the population of *Cactoblastis* decreased. Prickly pear began to recover at this stage and to regain lost territory, but the moth followed suit and prickly pear now exists only as scattered plants in the areas where it was once dense.

An interesting recent development of the biological control idea has been the use of sterile individuals in a population. An example of this method is provided by the screw worm fly of North America. The larvae of this insect, particularly in Texas, invade wound areas of cattle, and feed upon surrounding tissues. This enlarges the wound and encourages further egg laying by female flies. Until 1962 approximately 50,000 head of cattle were lost each year in Texas as a result of the activities of this pest. The expanse of the Texas ranges made regular examination and chemical treatment of beasts an impossibility, and the feasibility of a biological control measure was considered. Study of the life-cycle and population ecology of the insect provided a means by which this could be achieved. It was found that the female flies mated only once, and were then capable of egg laying. It was also found that flies could be rendered sterile, but otherwise unharmed, by irradiating them at the pupal stage of development. If male flies were sterilized and then released into the natural population, they competed with fertile males for the female flies and a proportion of ineffectual matings resulted. The more sterilized males released, the fewer fertilized females resulted in the natural population. Flies were bred in factories on enormous scales, sterilized and then released. Below are the figures of annual deaths of cattle from screw worm fly infestation:

1962—over 50,000
1963—over 5,000
1964—over 226
1965— 0.

Continuous invasion of new populations from Mexico has necessitated the continued release of the sterilized males to prevent a regrowth in screw worm fly populations.

These examples demonstrate that the control of man's environment need not be a clumsy and haphazard affair. The factor which differentiates these situations from the numerous occasions upon which man's attempts at environmental control have ended in tragedy, is the degree of research and planning which has preceded the introduction of the control measures in these cases. On both occasions any crude chemical control measures attempted proved ineffective, and thorough planning was unavoidable. The time is rapidly approaching when a long term view must be taken and crude, short-term remedies abandoned even where they would meet with apparent success.

Whilst considering the distributions of animals and plants it must have become evident that man himself is not above, nor is he outside, the problems which these distributions pose. As an animal with a physiology not unlike that of other animals, he too is dependent upon the physical factors of the environment for his continued existence. His cultural development has granted him a degree of insulation from his environment which has enabled him to invade areas of the globe which are physically inhospitable, but his insulation is not such that he can continue to exploit the world's resources and pollute its surface without any concern for the consequences. Nor is it such that he can continue to expand his population indefinitely: there comes a saturation point. The conservation of the environment ultimately means the conservation of mankind. It therefore demands our attention.

If we were to ask the question, what is man's greatest need, then the answer, as ever, must be knowledge, followed by the wisdom to act upon that knowledge. Knowledge of how our environment works involves research, and research is expensive. Action to conserve the environment based upon that research would cost very much more money. The dilemma in which man finds himself therefore is essentially a financial one. He must consider whether short-term economic gain and an artificially high standard of living for our present generation is worth the sacrifice of posterity.

References

1 TREVELYAN G.M. (1942) English Social History. Longmans, London.
2 BOLIN B. (1970) The carbon cycle. Sci. Amer. 223 (3), 124.
3 HAWKSWORTH D.L. & ROSE F. (1970) Qualitative scale for estimating sulphur dioxide air pollution in England and Wales using epiphytic lichens. Nature, Lond. 227, 145.
4 ARVILL R. (1967) Man and Environment. Penguin Books, Harmondsworth.
5 MELLANBY K. (1967) Pesticides and Pollution. Collins, London.
6 CLOUDSLEY-THOMPSON J.L. (1967) Animal Twilight. G.T. Foulis, London.
7 DODD A.P. (1940) The Biological Campaign against Prickly-Pear. Commonwealth Prickly-Pear Board.
8 GILBERT O. L. (1970) New tasks for lowly plants. New Scientist, 7th May, p. 288.

INDEX